Cognitive Systems Monographs

Volume 9

Editors: Rüdiger Dillmann · Yoshihiko Nakamura · Stefan Schaal · David Vernon

Heiko Hamann

Space-Time Continuous Models of Swarm Robotic Systems

Supporting Global-to-Local Programming

Springer

Rüdiger Dillmann, University of Karlsruhe, Faculty of Informatics, Institute of Anthropomatics, Humanoids and Intelligence Systems Laboratories, Kaiserstr. 12, 76131 Karlsruhe, Germany

Yoshihiko Nakamura, Tokyo University Fac. Engineering, Dept. Mechano-Informatics, 7-3-1 Hongo, Bukyo-ku Tokyo, 113-8656, Japan

Stefan Schaal, University of Southern California, Department Computer Science, Computational Learning & Motor Control Lab., Los Angeles, CA 90089-2905, USA

David Vernon, Khalifa University Department of Computer Engineering, PO Box 573, Sharjah, United Arab Emirates

Editors

Dr.-Ing. Heiko Hamann
University of Graz
Department of Zoology
Universitätsplatz 2/I
8010 Graz
Austria
E-mail: heiko.hamann@uni-graz.at

ISBN 978-3-642-13376-3 e-ISBN 978-3-642-13377-0

DOI 10.1007/978-3-642-13377-0

Cognitive Systems Monographs ISSN 1867-4925

Library of Congress Control Number: 2010928224

Typeset & Cover Design: Scientific Publishing Services Pvt. Ltd., Chennai, India.

Printed on acid-free paper

5 4 3 2 1 0

springer.com

Abstract

In this work, a generic model in as far as possible mathematical closed-form is developed that predicts the behavior of large self-organizing robot groups (robot swarms) based on their control algorithm. In addition, an extensive subsumption of the relatively young and distinctive interdisciplinary research field of swarm robotics is emphasized. The connection to many related fields is highlighted and the concepts and methods borrowed from these fields are described shortly.

Large groups of small robots, mostly of limited equipment are applied in swarm robotics forming a decentral system. All robots are autonomous and act on the basis of locally available information. The development of the control algorithm, that is to be executed locally on each robot, has proven to be difficult. This development of the local control algorithm is defined and constrained by the global task (global-to-local programming, or also micro-macro problem). The classical reductionistic approach is of limited use here (problem of designing emergence). For example, the resulting behavior of the robot swarm often contradicts the intuition of the program developer due to effects of the many robot–robot interactions that cannot be anticipated.

The support of the swarm algorithm developer by models is an approach that has already been discussed in the literature several times. The quickly available predictions of the model are supposed to support the development early before the implementation on the robots. Even complete parameter intervals can be scanned for optimal values. Furthermore, the development and the application of models can result in a better understanding of the effective processes in swarms concerning both the general understanding and in a particular application. The modeling approach proposed in this work is particularly distinguished by the explicit representation of space and the, at least partially existent, formal connection between the micro- and the macro level.

The basic model of the robot positions is motivated by Brownian motion and consists of a pair of corresponding equations. While the Langevin equation (a stochastic differential equation) gives a local (microscopic) description of concrete trajectories, the Fokker–Planck equation (also Kolmogorov forward equation, a partial differential equation), that can be analytically derived from the Langevin equation, gives a global (macroscopic) description by means of probability densities. This physical model was extended to a generic model of communicating robot groups based on heuristic arguments. This model approach has a variety of applications, however, the adaptation to a specific control algorithm is a demanding modeling step.

The proposed model is validated against several swarm robotic scenarios applying real robots and simulations: collision-based adaptive aggregation, collective perception, collective phototaxis, foraging with virtual pheromones, and tree-like aggregation. The required adaptation of the model to the according situation is exemplified (modeling state transitions, parameter selection, measurement etc.). The achieved accuracy of the model predictions is good and sufficient to be a support in the algorithm development phase.

Contents

Chapter 1
Introduction and Purpose

The context of this work is the young field of swarm robotics. A preeminent characteristic of swarm robotics is its inter-disciplinarity. Diverse disciplines such as philosophy or engineering are relevant. Classical robotics is already an interdisciplinary field uniting mechanical engineering, electrical engineering, cognitive systems, artificial intelligence, and even philosophy. However, swarm robotics is even more extensive including fields of biology, classical chemistry, and statistical physics.

Interdisciplinarity is both a blessing and a curse: On the one hand, it implies diversity and richness making a research project deep and interesting. On the other hand, it might imply the relevance of a vast number of fields making the research unclear and all efforts will inherently be incomplete. In this work we try to integrate the relevant aspects of swarm robotics into the context of science.

A detailed description of swarm robotics will be given later, now we just cite the definition of Şahin and Spears (2005): "The study of how a swarm of relatively simple physically embodied agents can be constructed to collectively accomplish tasks that are beyond the capabilities of a single one." Due to its novelty swarm robotics is motivated by its potential future applications. One important class of scenarios is the application in hardly accessible areas (e.g., pipe systems) because of the robot's small overall size. Another benefit of the swarm systems is the high degree of robustness, that is, the system's effectiveness is ensured, even if a high percentage of the swarm is lost. This robustness results in a distribution of risk which is an appropriate approach to safety critical scenarios. However, before swarm robotic systems can successfully be applied, an efficient way to design and implement them needs to be found. Unfortunately, the design of the control algorithm, that is locally executed on the individual robot, proved to be difficult.

Multi-robot systems are non-linear systems (c.f. Section 2.5 on page 18). Therefore, they might show complex behavior. An additional complexity originates in the high number of robot-robot interactions whose effects can, in general, barely be estimated. This characteristic is shared with many other systems such as ant colonies, the laser, or cities. These systems show features that are summarized by calling them *self-organized*. Self-organization is basically the formation of structure out of disorder within a system without any regulative influence from the outside. Many of

H. Hamann: Space-Time Cont. Models of Swarm Rob. Sys., COSMOS 9, pp. 1–3.
springerlink.com

these systems show so-called *emergent* properties as well. Instead of defining emergence here we just cite Johnson (2001), who discusses the process of maturing in an ant colony that takes longer than the life of an individual ant: "How does the whole develop a life cycle when the parts are so short-lived? It would not be wrong to say that understanding emergence begins with unraveling this puzzle." With regard to the design of systems, that show emergent properties, Johnson (2001) summerizes Gerald Edelman's experience: "You never really know what lies on the other end of a phase transition until you press play and find out. That is the lesson of Gerald Edelman's recipe for simulating a flesh-and-blood organism: you set up a system of various pattern-recognition devices and feedback loops, connecting the virtual organism to a simulated environment. And then you see what happens." This potential unpredictability of the system properties is exactly the challenge we face when trying to design the control algorithm of a swarm robot.

As the classical programming approach for robot swarms has proved to be a difficult task researchers are searching for alternatives. Besides the application of evolutionary or learning algorithms the model approach is one option. The basic idea is to support the algorithm designer by providing a tool that predicts the behavior defined by the current algorithm as precisely as possible and within a short period of time. The purpose of this work is not the synthesis of algorithms or the online application of the model (i.e., not on the robots or simultaneous with the running robots) but sufficient offline predictions of the swarm behavior calculated on up-to-date standard desktop PCs.

On the one hand, the problem of programming robot swarms has exposed challenges that had not to be solved by engineers and computer scientists before. On the other hand, these challenges are well known to philosophers since many years but these challenges have not been in connection to such practical fields as engineering. By trying to design self-organizing systems as well as systems showing emergent behavior we tackle a whole bunch of interdisciplinary problems. Thus, the models presented in this work do not only apply to swarm robotics but might also prove to be a suitable tool for biologists or might be a tool to come to a better understanding of what actually emergence is.

1.1 Objectives

The main goal of this work is to develop a model of robot swarms that supports the control algorithm designer. This is guaranteed by a model that:

- reliably delivers sufficiently accurate predictions of the overall swarm behavior that is to be expected in the average based on the current control algorithm,
- is generically applicable to typical swarm robotic scenarios,
- establishes an explicit connection between the level of the individual (micro-level) and the level of the swarm (macro-level),
- is capable of representing inhomogeneous space (e.g., inhomogeneous distributions of the robots over the arena, obstacles, regions of different conditions in the arena, etc.) in a straightforward manner,

- is computationally efficient and has a computational complexity independent of the swarm size,
- can be presented concisely and that allows, therefore, easy comparability.

1.2 Approach

The approach that is pursued in this work to achieve the objectives is based on the physical model of Brownian motion. Out of a stochastic differential equation (Langevin equation) a partial differential equation (Fokker–Planck equation) is derived. Both equations are space-time continuous, guaranteeing the representation of inhomogeneous space, and the derivation delivers the explicit connection between the micro- and the macro-level. This model is extended to a generic model of communicating robot groups that is adapted in a final modeling step to the considered swarm scenario. The computational efficiency is ensured by efficient numerical techniques, the main dependence on the spatial discretization, and the independence of the swarm size. The interactions between robots are modeled explicitly, if these interactions are analytically describable by standard models of physical phenomena (e.g., diffusion processes, fluid dynamics). The interactions are modeled by heuristic approaches, if such models based on probability densities do not exist for this type of interaction (e.g., point-to-point communication). High accuracy is achievable but only by an overhead of heuristic parameters that need to be selected and measured in smaller and simpler scenarios than the actual considered scenario. In comparison, predictions, that are only qualitatively correct, are much easier to obtain concerning the resources that are needed to be invested during the model development.

1.3 Outline

In Chapter 2 we define fundamental concepts of swarm robotics and give an overview of related research fields and crucial issues. In Chapter 3 we summarize the state-of-the-art in modeling swarm/group/multi-particle systems and discuss the state-of-the-art in designing emergent behavior. The model itself is motivated, introduced, and discussed in Chapter 4 including a discussion of interpreting swarm dynamics as computation. In Chapter 5 we validate our model and the "swarm dynamics is computation"-interpretation in five scenarios: collision-based aggregation, collective perception, collective phototaxis, foraging using virtual pheromones, and tree-like aggregation.

Chapter 2
Fundamentals of Swarm Robotics – An Interdisciplinary Approach

Yet such is oft the course of deeds
that move the wheels of the world:
small hands do them because they must,
while the eyes of the great are elsewhere.

J.R.R. Tolkien in "Lord of the Rings"

We give definitions of important terms and short introductions to related fields of research.

2.1 Definition of Fundamental Concepts

In the following we give definitions for concepts that are fundamental in swarm robotics. These concepts will be used throughout this work.

2.1.1 Agent

An *agent* is an autonomous entity which perceives through sensors and acts through actuators as defined by Russell and Norvig (1995). They group agents into five classes based on their degree of perceived intelligence and capability. Here, only the first two classes are relevant. The first class is the simple reflex agent which acts based on the current perception only and, thus, without any use of memory. The second class is the model-based reflex agent which manages an internal state as a model of the environment. The agents in this work are reflex agents. Most of them make use of memory. However, they have an internal state that can hardly be called a model of the environment. Typically, they just store their internal state comparable to a finite state automaton, sometimes combined with a floating point variable representing a "virtual feature of the environment" (e.g., value of a virtual potential

H. Hamann: Space-Time Cont. Models of Swarm Rob. Sys., COSMOS 9, pp. 5–26.
springerlink.com

field). Thus, the considered agents are hybrids between the simple reflex agent and the model-based reflex agent. Not considered are tele-operated or semi-autonomous robots. In this work an agent will always be a real or simulated robot, thus, we use agent and robot as synonyms. In principal, animals, for example, bees or ants, can also be such an agent which is not excluded.

2.1.2 Agent–Agent Interaction

An *agent–agent interaction* is basically the encounter of two robots with mutual influence. In a simple example, such an interaction occurs when robots A and B recognize each other as obstacles and they rotate away from each other to avoid a collision. This is called collision avoidance behavior. If only robot A perceives B this encounter could also be an interaction but with influence to robot A only. This is an inadequacy of the literal sense of "interaction" as B's contribution is only passive by entering the sensor range of robot A. For simplicity, the mere encounter of two robots will be called collision in the following.

2.1.3 Macroscopic and Microscopic Level

The *microscopic* level (or micro-level) of a multi-agent system is the individual level. In a microscopic description of a multi-agent system the individual agent is addressed, that is, we can determine, for example, trajectories of agents and distances between agents.

The *macroscopic* level (or macro-level) of a multi-agent system is the group level. Individual agents are not addressed as we work only on group fractions or densities of agents, that is, trajectories of individual agents cannot be determined, only, for example, a mean trajectory of the swarm's barycenter.

A level in between is also conceivably even though of less relevance. This intermediate level can be called *mesoscopic*. In a mesoscopic description neither individual robots are represented nor do we have a fully abstract model of agent densities. An example is the Boltzmann equation in physics, which models agent densities but also agent velocities and collisions.

2.1.4 Phenomenological Approach

In this work the adjective *phenomenological* will be used without a deeper philosophical meaning in the sense of phenomenology (Moran, 2000). In the natural sciences it is rather used to describe preliminary approaches that just describe phenomena instead of applying a profound theory or interpreting the observed features. For example, a phenomenological approach to swarm behavior is to develop a macroscopic model based on the observed swarm-level features that describes the swarm behavior. However, such a model will not have much explanatory power without any connection to the underlying microscopic rules. Say, we assume that the macro-behavior is defined by the micro-behavior. Note that this assumption will be challenged in Section 2.1.6 where the concept of emergence is discussed. Then it follows that a phenomenological model cannot explain the swarm behavior as it only addresses macroscopic features, that is, it only describes the effects of the microscopic level.

2.1.5 Self-organization

A system is *self-organizing*, if it generates structures by internal organization without any guiding influence from the outside. The four principles of self-organization are (Camazine et al., 2001):

1. positive feedback,
2. negative feedback,
3. amplification of fluctuations,
4. multiple interactions.

Positive feedbacks ensure the intensive utilization of developed solutions or desired behavior. Negative feedbacks control the behavior to avoid inefficient processes and deadlocks. Fluctuations caused by random behavior of single agents might be amplified, for example by attracting other agents, and lead to new solutions and exploitation. The high number of agent–agent interactions generates a high degree of correlation between the agents that allows the emergence of macroscopic structures.

Examples for self-organizing systems are: spontaneous magnetization, the laser (Haken, 1971), percolation (Grimmett, 1999), diffusion-limited aggregation (Tom A. Witten and Sander, 1981), and ant colonies (Camazine et al., 2001; Deneubourg et al., 1990). Properties that are typically ascribed to self-organizing systems are: increase in order, autonomy, adaptability, robustness, and dynamics (Wolf and Holvoet, 2005). Self-organization is different from the concept of emergence that is discussed in the next section. A self-organizing system does not necessarily have to create novel properties but many self-organizing systems display emergence.

2.1.6 Emergence

Emergence is both an intriguing subject in swarm robotics and a rather vague concept. This ambivalence is caused by the promising opportunities it might offer ("getting something out of almost nothing", see Section 2.6.1 on page 21) and the difficulty to capture and to define it. It is the concept of emergence that links engineering with fundamental philosophical concepts such as holism or even metaphysics. Thus, a critical inspection of this topic by the engineer is a well considered way. However, ignorance would be questionable because emergence might offer extensive possibilities, for example, the construction of complex systems might be possible by an unproportionally small effort — the engineers dream.

A full definition of emergence is not given here since finding a well definition is a research project of its own and many researchers have already tried to define it (Deguet et al., 2006; Abbott, 2004; Weinberg, 1995; Kubík, 2003; Wolf and Holvoet, 2005; Kubík, 2001; Darley, 1994; Bonabeau, 2002). Others do not even try to define emergence: "It is unlikely that a topic as complicated as emergence will submit meekly to a concise definition, and I have no such definition to offer" (Holland, 1998). It is still a nondistinctive concept as stated by Holland (1998): "Despite its ubiquity and importance, emergence is an enigmatic, recondite topic, more wondered at than analyzed." However, to allow the use of the word "emergence" and the discussion of the concept we give a simplified and preliminary definition: A system has *emergent* properties (and is then called emergent), if a behavior on the macro-level is not explicitly programmed on the micro-level. This is a careful and maybe even inoffensive definition but it communicates a suggestive meaning for systems that are labeled as "emergent".

Emergent behavior is an important ingredient of Swarm Intelligence according to Dorigo et al. (2000): "Solutions to problems faced by a colony are emergent rather than predefined." In swarm robotics there is an emphasis on emergence as well following Dorigo and Şahin (2004):

> Swarm robotics consists in the application of swarm intelligence to the control of robotic swarms, emphasizing decentralization of the control, limited communication abilities among robots, use of local information, emergence of global behavior and robustness.

However, there is no consensus reached on a definition of emergence as, for example, stated by Bayindir and Şahin (2007):

> Although lots of efforts spent for understanding how emergence occurs, there is no satisfactory theory explaining what characterizes emergence or what are the conditions for its existence.

Nevertheless, swarm roboticists have spared no effort to identify their own point of view and to give definitions, for example, Martinoli (1999):

> In other words, we can speak of emergent individual behavior if the resulting robot behavior was not explicitly programmed in any of its functional blocks and arises from interactions among them [...] and with the environment.

or Bayindir and Şahin (2007):

> Emergence is a key property in complex systems which means the behavior of the complex system cannot be understood by examining only the components of the system. Although the components of the complex system can be simple, the resultant system may be complex because of the interactions of the system components.

Additionally, following Beni (2005) and Şahin (2005), the concept of emergence serves as an indicator that an algorithm is consistent to the principles of swarm robotics, for example, Bjerknes et al. (2007):

> The algorithm meets the criteria for swarm robotics [...]. We have a highly robust and scalable swarm of homogeneous and relatively incapable robots with only local sensing and communication capabilities, in which the required swarm behaviors are truly emergent.

A more in-depth discussion of this concept and its philosophical background is given in section 2.6.1 on page 20.

2.1.7 *Stigmergy*

Stigmergy is an artificial term created by Grassé (1959, 1967). A stigmergic system has a process that changes the environment, and the system itself reacts again to the changes in the environment. Features of the environment serve as stimuli to the system's behavior. This delayed feedback loop might organize the interactions between the environment and the system in a way that creates an ordered state. In natural swarms, stigmergy is often the driving force behind phenomena, such as ant trails, termite mounds, or wax combs. Therefore, stigmergy is an important research subject in swarm robotics, for example, in puck sorting (Bonabeau et al., 1999) or artificial pheromone trails (Payton et al., 2001; Sugawara et al., 2004; Hamann et al., 2007).

2.1.8 *Micro-Macro Link*

The *micro-macro link* or also called *micro-macro problem* is a technical term in sociology (Alexander et al., 1987; Schillo et al., 2000). This term is based on the concept of micro-interactions (interactions between individual humans, for example, actions that are aimed to change the actions of others) and macro-structures (structural rules defined, for example, by the human society). The micro-macro link describes the mutual influence of the macro-structure to micro-interactions and vice versa. The action of an individual might be influenced or caused by the individual's perception of the macro-structure. This behavior influences, in turn, the macro-structure leading to a closed loop. Due to this loop structure it is difficult to distinguish whether the cause of a considered action is found on the micro- or the macro-level as indicated by the term "micro-macro problem".

This concept is related to self-organization and emergence in swarms. The individual perceives parts of global structures (e.g., the gradient of a pheromone trail leading to food), reacts to these (e.g., turning to the food), and influences them (e.g., by dropping pheromones). Again, it is difficult to determine the macro-behavior based on the micro-behavior and vice versa.

2.2 Swarm Intelligence

Swarm intelligence is a subfield of artificial intelligence and focuses on decentralized collective behavior in large multi-agent systems (swarms) showing self-organization. Beni and Wang (1989) introduced this term discussing cellular robotic systems. Millonas (1994) defines five basic properties of "swarm intelligent" systems, cf. Kennedy and Eberhart (2001) and Bonabeau et al. (1999):

- Proximity principle: The agents execute simple computations concerning space and time.
- Quality principle: The agents respond to quality factors, such as determining the safety of a location.
- Principle of diverse response: The agents distribute themselves and their resources in a variety of ways instead of concentrating on too narrow behavior.
- Principle of stability: The swarm is stable against repetitive fluctuations in the environment and does not oscillate.
- Principle of adaptability: The swarm is sensitive to changes in the environment that require a change in the swarm behavior.

Furthermore, it is commonly assumed that a decentralized collective behavior is based on local information only and global communication is typically forbidden, which we will call the principle of local information.

2.2.1 From Natural to Artificial Swarms

The interest in artificial swarms was initiated by amazingly simple models of swarm behavior by Reynolds (1987). He presented three simple rules that suffice to generate the behavior of swarms, flocks, herds, and schools in simulations:

- Separation: Move away from other agents that get too close.
- Alignment: Move in the same direction as the neighboring agents.
- Cohesion: Move towards the barycenter of neighboring agents.

These rules are defined on the microscopic level. They generate self-organizing behavior because the system is ordered (cluster of agents with similar directions) without any influence from the outside. The behavior is also emergent because the collective motion is a novelty that cannot be observed on the micro-level. Another interpretation is that the aggregation behavior is not explicitly programmed on the micro-level. This statement is, however, indefinite and reflects the inconclusive character of emergence as a concept.

Self-organization is a common feature in biological systems on all levels: processes in the animal, animal behavior, and swarm behavior (Camazine et al., 2001). The step from this natural self-organized behavior via biological models (see Section 3.1.6 on page 32) to the design of artificial swarms is small because once the behavior is artificially reproduced it can also be adapted and designed for desired tasks. By designing artificial swarm behavior swarms become problem solvers.

2.2.2 Swarms as Problem Solvers

The prime example for an application of swarms as problem solvers is the metaheuristic ant colony optimization (ACO) proposed by Dorigo and Caro (1999). ACO is inspired by ants laying down pheromone trails. This behavior is broadly resembled by virtual ants that “walk” through a graph and develop pheromone trails. ACO was applied to the traveling salesman problem (TSP) and produced near-optimal solutions. Furthermore, it is distinguished by its high degree of adaptability to changes in the graph topology, just as natural swarm systems. That is why it was successfully applied to real-time problems.

A related meta-heuristic is particle swarm optimization (PSO) introduced by Kennedy and Eberhart (1995). PSO resembles vaguely social learning by agents that move through the solution space telling each other where good solution are situated. However, PSO uses a global best value that is known by all agents and therefore violates the principle of local information.

Besides these software swarms all swarm robotic implementations can be interpreted as artificial swarms in hardware, see Section 2.3.3 on the next page.

2.3 Robotics

2.3.1 Mobile Robotics

Mobile robotics can arguably be dated back to 1515, when Leonardo da Vinci presented a mechanical lion at the arrival of King Francis in Lyon. However, the sources are not definite and the ascription to Leonardo might be a mystification. This lion possibly was nothing more than an automaton based on a clockwork but able to move.

Sometimes the origins are dated back to World War II, where so-called "smart bombs" with guiding systems were applied. At least in the 1940s a variety of projects began developing wheeled robots.

Today we have a huge variety of autonomous mobile robots that are legged, wheeled, or have tracks; but there are also unmanned aerial vehicles and autonomous underwater vehicles. We ignore this vast variety here and focus on robots of small overall size instead.

2.3.2 Micro-Robotics

The word "micro" in micro-robotics (also called microbotics) has two meanings here. First, it indicates the small size of the robots, which is today, however, far above micrometers and rather in the millimeter or centimeter range. Second, it indicates the high path accuracy that can actually go down to nanometer-scale ($10nm$ in case of MINIMAN, see below). The connection to swarm robotics is drawn because a small overall size of swarm robots is at least beneficial if not necessary. Otherwise, it would barely be feasible to deploy many dozens or even hundreds of robots under laboratory conditions. Micro-robotics served as the stepping stone to swarm robotics at the Institute for Process Control and Robotics at which this work was developed (Fatikow and Rembold, 1993; Fatikow et al., 2000; Wörn et al., 2000). Exemplarily, we want to highlight the MINIMAN, the MiCRoN, and the I-SWARM project. These projects show the transition from just building small robots to small groups of small robots, and finally to large groups or swarms of small robots.

According to the official statements (Fahlbusch et al., 1999)

> the main idea of the MINIMAN project was the development of a smart micro-robot with five degrees of freedom and a size of a few cubic centimeters, capable of moving and manipulating by the use of tube-shaped and multilayered piezo-actuators. Controlled by visual and force/tactile sensor information, the micro-robot is able to perform manipulations with a motion resolution down to $10nm$ in a tele-manipulated or semi-automated mode, and so it frees humans from the tedious task of having to handle minuscule objects directly. Equipped with micro-machined grippers, the robot takes over high-precise grasping, transport, manipulation and positioning of mechanical or biological micro-objects, under a light microscope or within the vacuum chamber of a scanning electron microscope. A powerful computer system using inexpensive PC-compatible hardware components will ensure the robot operation in real-time.

The robot is shown in Figures 2.1 on the facing page and 2.2 on the next page.

Again citing the official statements (Brufau et al., 2005)

> the goal of the MiCRoN Project was to develop a system that is based on a cluster (five to ten) of small (in the size of a few cm^3) mobile autonomous robots. These wireless micro-robots, each equipped with on-board electronics for control and communication, cooperate to accomplish a range of tasks associated with assembly and processing from the nano- to the micro-range.

A prototype of the robot is shown in Figure 2.3 on the facing page in comparison to MINIMAN 3.2.

In the I-SWARM project the goal was to produce a "true" (robotic) self-organizing swarm concerning the size of the swarm and the size of the individual robots (Seyfried et al., 2005). The idea was to establish a mass production for autonomous robots making, in principle, the production of numbers as high as 10^3 feasible. The robot itself has a size of just $3mm \times 3mm \times 3mm$, is equipped with four infrared emitters and sensors, three vibrating legs, and a solar cell as energy supply as well as additional sensor (see Figure 2.4 on page 14). On the software side, control algorithms were developed and implemented that lead to self-organization and emergent properties. For example, the algorithms analyzed in Sections 5.1 on page 77 and 5.2 on page 89 were developed within the I-SWARM project. The project can be classified as belonging to the field of nature-inspired minimalist swarm robotics (cf. Section 2.3.3).

2.3.3 *Swarm Robotics*

The origin of swarm robotics seems to be swarm intelligence as documented by Beni (2005). Thus, swarm robotics can be interpreted as swarm intelligence applied to robotics. A short definition is given by Şahin and Spears (2005): "The study of how a swarm of relatively simple physically embodied agents can be constructed to collectively accomplish tasks that are beyond the capabilities of a single one." The idea is to build rather many small and inexpensive robots than building only a few, big, and expensive robots. The group of small robots is supposed to accomplish the same task as a single complex robot or a small group of complex robots.

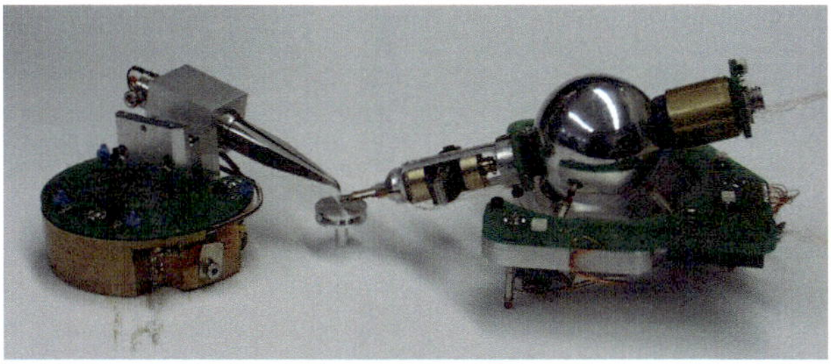

Fig. 2.1 The MINIMAN 4 robot on the left, and the MINIMAN 3.2 on the right

Fig. 2.2 MINIMAN 3.2 with Moiré marks, view from the top

Fig. 2.3 Prototype of the MiCRoN robot in comparison to MINIMAN 3.2

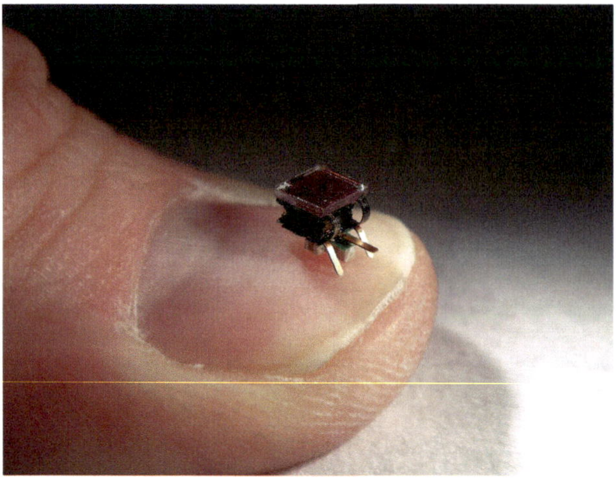

Fig. 2.4 The I-SWARM robot on an adult's thumbnail

The first main benefit is robustness, that is, the loss of some robots will not cause a breakdown of the system (Şahin, 2005). This can be interpreted as a spread of risks. For example, one can invest money in a single complex space probe. A single failure that leads to the loss of this probe will make the whole project a failure. If the same amount of money is invested in hundreds of small space probes, that have in cooperation the same capabilities as the complex probe, a loss of some of them might be absorbed. A second benefit is flexibility. The hardware of these robots is not necessarily custom-built for a given task. In cooperating and grouping differently the robot swarm might accomplish other tasks as well while the complex robot would need a hardware reconfiguration or a redesign. The third benefit is scalability. A swarm robotic algorithm can be applied unchanged to a group of any (reasonable) size due to relying on local information only. Any global communication would prohibit scalability and is avoided.

However, there is some confusion about what swarm robotics exactly is as identified by Sharkey (2007). The minimalist approach (primarily concerning the hardware) seems not to be a consensus in the community anymore. Therefore, Sharkey defines three sub-areas of swarm robotics:

1. scalable swarm robotics: not minimalist and not directly nature-inspired,
2. practical minimalist swarm robotics: not directly nature-inspired,
3. nature-inspired minimalist swarm robotics.

Other terms are applied as well, citing from Sharkey (2007):

> [Swarm robotics] is one of a number of related terms that have been applied to multi-robot systems: others include "minimalist robotics", (e.g., Wilson et al. (2004)), "robot colonies" and "distributed robotics" (Dorigo and Şahin, 2004), and "large-scale minimalist multi-robot systems" (Jones and Matarić, 2003), whilst Kube and Bonabeau (2000) suggest that both "swarm-based" and "collective robotics" describe the same

approach. The field is still undergoing rapid development, and as yet there is not a clear consensus about which terms should be used, or what their defining features are.

From this point of view this work is mainly nature-inspired minimalist swarm robotics which can also be called nature-inspired minimalist collective robotics.

The swarm robotic hardware, that is more than just a case study and still in use, can be counted on the fingers of one hand: swarm-bot — also known as s-bot (Dorigo et al., 2006), Jasmine (Jasmine (2008); Kornienko et al. (2005a), see also Section 5.1.1 on page 78), alice (Caprari et al., 1998), ePuck (ePuck, 2008), and "The Swarm" by James McLurkin (unpublished). Furthermore, there are developments of flying swarms, see for example Nardi and Holland (2007), and underwater swarms, see Kottege and Zimmer (2008).

In contrast to the few hardware, there are many software-related studies on swarm strategies and swarm algorithms: virtual pheromones (Payton et al., 2001) or artificial physics (Spears and Gordon, 1999) as control technique, trail following (Vaughan et al., 2000; Hamann et al., 2007), task allocation in a swarm (Jones and Matarić, 2003; Lerman et al., 2006), inspection of mechanical devices (Correll et al., 2006; Correll, 2007), physical interaction between robots (Mondada et al., 2005), coherent swarming and taxis (Nembrini et al., 2002; Bjerknes et al., 2007), collective perception (Schmickl and Crailsheim, 2006), and a special operating system for swarm robots (Szymanski and Wörn, 2007).

2.3.4 Embodied Cognitive Science

Embodied cognitive science as postulated by Pfeifer and Bongard (2006) is also connected to swarm robotics. This concept aims primarily at creating artificial intelligence and claims that true artificial intelligence is only achievable by a holistic approach unifying body and mind, that is, by robotics. For example, the legs of a robot are designed in a way simplifying radically the control mechanism needed to make it walk (Iida, 2005). The connection to swarm robotics is the focus on embodied agents, just as swarm robotics is the science of embodied swarm intelligence. In addition, embodied cognitive science is also related to the minimalist approach of swarm robotics as the robots are also reduced in their complexity and the physical presence of neighboring robots, that is perceived through the sensors, is essential for swarm behavior.

2.3.5 Sensor/Actuator Networks

From the sensor/actuator network point of view the robot swarm is a group of mobile, autonomous, wireless sensor nodes provided with tools to manipulate the environment. The two important components of a sensor node are the sensor (at least one) and a communication device which is wireless and typically radio communication. However, other communication principles such as light (e.g., infrared) or sound might be applied as well. The sensor is usually powered by a battery. Therefore, energy consumption is crucial and communication overhead is tried to be traded for computation overhead; but the computational power is also limited because micro-controllers with low performance are used.

The main focus of sensor network research and the related field of ad hoc networks is on communication. Communication architectures and protocols need to be defined and analyzed in order to apply them to spatially distributed networks of a possibly dynamic topology. Challenges are the design for robustness on the network-level and scalability because individual nodes might be unreliable and huge networks (10^3 and more nodes) should be supported. Due to the lack of large-scale implementations evaluations are typically done with simulators such as ns-2 (McCanne et al., 1997).

Global communication, for example, point-to-point communication through the whole network, is not prohibited a priori. However, global communication is in general not scalable. Also hierarchical approaches are only a compromise because they jeopardize the robustness of the network. Hence, sensor network techniques, that are strictly designed for scalability, might be applied to swarm robotics systems. Generally, this will be in the field of scalable swarm robotics (not minimalist and not directly nature-inspired, cf. Section 2.3.3 on page 14). This knowledge transfer has just begun using inhomogeneous swarms out of swarm robots and sensor nodes (Correll et al., 2006).

2.4 Software Concepts for Distributed Systems

2.4.1 (Distributed) Artificial Intelligence and Multi-robot Systems

Robotics might be considered as the supreme discipline of artificial intelligence — assuming the more than ambitious goal of building a robot with all the abilities of a human being including: movement in a variety of environments, image/speech recognition, grasping of a variety of objects, solving tasks widely assumed to require intelligence, and so on. While all of these challenges can be investigated by using a single robot the field of distributed artificial intelligence investigates the possibilities of cooperation in multi-robot systems. One of the most popular challenges of this field is robot soccer (Robocup). Although general conceptual delineation to swarm robotics might seem to be hardly possible it is often rather simple in precise scenarios. Multi-robot systems assuming global communication should not be considered swarm robotic systems. Also high complexity of the involved robots might be an indicator that the underlying concepts are not based on swarm intelligence. However, this is only a true delineation for minimalistic swarm robotics.

2.4.2 Multi-Agent Systems

A multi-agent system is a system consisting of several agents (in the sense of Section 2.1.1 on page 5). The main focus of this field is on software agents for tasks such as online trading or information search (Weiss, 1999; Woolridge, 2001). A second focus is on agent-based modeling, for example, with applications to sociology (Conte et al., 2001; Schillo et al., 2000). Furthermore, multi-agent systems serve as models for robotic systems as, for example, in the robot soccer domain (Vetulani, 2002). This is the connection to swarm robotics. On the one hand, however, many studies in multi-agent systems assume global communication similar to sensor networks. Thus, scalability is again not ensured. On the other hand, there is also research on scalable decentral systems. In fact, drawing the borderline between multi-agent systems and software simulations of swarm robotic systems is challenging because the transition from large decentral multi-agent systems to swarms is blurred. This is true, especially, for scalable swarm robotics (cf. Section 2.3.3 on page 12). Thus, the software and the modeling subfield of swarm robotics can be regarded as multi-agent systems with emphasis on scalability and biological inspiration. Multi-agent systems are at least the historical basis for swarm robotics, just as they are for swarm intelligence.

2.4.3 Artificial Life

Founded by Langton (1989a), artificial life (or Alife) is the study of life itself and processes related to life, especially evolution. There is a vast variety ranging from abstract approaches based on models and simulations to actual "wet" experiments in biochemistry. Additionally, there are also studies using robots or studies with relevance to robotics, such as the Braitenberg vehicle (Braitenberg, 1984), which is commonly and retroactively assigned to artificial life.

Artificial life and swarm robotics have the biological inspiration in common, which is intrinsic for Alife. Furthermore, there are simulations of large-scale agent groups in Alife. The difference between the two fields is the "observant mindset" in Alife, in contrast to the constructive purpose in swarm robotics. The swarm roboticist wants to design a system that solves a given problem. The Alife researcher strives for an insight of what life is and how it works. Thus, Alife research is typically rather phenomenological also due to the high complexity of their research subject.

Concerning the work at hand, it seems to be a hybrid of at least these two fields. On the one hand, it aims to increase the design efficency of swarm robotic systems. On the other hand, the proposed model might serve as a tool for new insights as well; for example, concerning the problem of emergence, which is also often called on and often discussed in Alife.

2.4.4 Amorphous Computing

Amorphous computing as a general concept was introduced by Abelson et al. (2000). The research of amorphous computing is focused on finding programming paradigms for a continuous "smart" medium or rather for a discrete approximation of such a medium which is composed of immobile particles. The approach is biologically inspired by the cooperation of cells in natural organisms, the hardware is assumed to be unreliable, and the network topology is assumed to be unknown. Obviously, amorphous computing is also connected to sensor/actuator networks. Essentially, it is research on programming languages that are defined on the continuous medium abstraction and the challenge is to find compilers that translate these abstract description into executable code for the particles (Yamins and Nagpal, 2008; Yamins, 2007, 2005).

2.4.5 *Self-awareness in Distributed Systems*

The concept of "self-awareness" in distributed systems, also called "adaptive information processing" in decentralized systems introduced by Mitchell (2005, 2006) is focusing on essentially the same principles as swarm intelligence but with a slightly different emphasis. Mitchell defines four principles:

1. Global information is encoded as statistics and dynamics of patterns over the systems components.
2. Randomness and probabilities are essential.
3. The system carries out a fine-grained, parallel search of possibilities.
4. The system exhibits a continual interplay of bottom-up and top-down processes.

The encoding of global information in patterns is relevant to swarm robotics and apparently underestimated as it is seldom refered to (see also Section 4.7 on page 68 and especially the quote of Litus). Randomness is also important in the movements of swarm robots as discussed in Section 4.4 on page 61. The "parallel search" corresponds to the swarm intelligent problem solvers (Section 2.2.2 on page 10) as well as to the exploration behavior of robot swarms. We omit a description of the rather vague concept of an "interplay of bottom-up and top-down processes".

2.5 Science of Self-organization

In order to design self-organizing systems showing emergent properties, a science of self-organization would be helpful. There seem to be only two theories that can rightly be called "theory of self-organization". One is the theory of dissipative structures founded by Prigogine (1967) and the otherone is synergetics founded by Haken (1971). Both originate in physics and chemistry and, thus are based on mature mathematical descriptions.

2.5.1 *Theory of Dissipative Structures*

A dissipative structure is a system that continuously imports free energy from the outside and exports entropy, for example, it looses heat, but is stable and shows macroscopic patterns (Nicolis and Prigogine, 1977; Prigogine, 1967, 1969). This behavior is observed in states that are far from the thermodynamic equilibrium. Therefore, the study of such systems is also called far-from-equilibrium thermodynamics.

A prominent physical example is convection in a liquid. For example, heating a shallow container of oil from below leads to a special phenomenon. At first, the heat is transfered to the top of the liquid by means of conduction. This system behaves linear: A small increase or decrease in the temperature of the heat source will lead to small changes in the system at any position. All molecules in the system, that is the micro-level, show random behavior. The macro-state is homogeneous.

However, with increasing heat the system switches to convection and dramatic changes are observed. Convection cells are formed, here: Bénard cells. The former random motion of the molecules is now transformed into ordered motion. The molecules ascend and fall forming the rotation of the cells. Neighboring cells rotate in the opposite direction – a spontaneous symmetry breaking. This dissipative structure is stabilized by an exchange of energy with the environment.

Dissipative Structures are a sound theory for physical and chemical systems. Many approaches were proposed to extend this theory to biological/multi-agent systems, see for exmaple Parunak and Brueckner (2001), and even to systems of higher complexity such as the human society. However, a complete theory is still pending.

2.5.2 Synergetics

Synergetics is the study of macroscopic systems consisting of many non-linearly interacting subsystems (Haken, 1971, 1977). It is based on the "order-parameter concept" (Ginzburg and Landau, 1950) which was generalized by Haken to the "slaving-principle". The idea is that microscopic and stable processes acting on a small time-scale (fast-relaxing) are determined by macroscopic and instable processes, the order parameters, acting on a large time-scale, that is, they are slowly varying and form macroscopic patterns. The order parameters are in competition with each other and are independent of the details on the micro-level. Thus, the macro-level is completely described by the order parameters.

This theory originates from thermodynamics and this field is also the one where it fits in the best way. For thermodynamic systems the slaving-principle describes self-organization, which is interpreted as the formation of order parameters leading to an increase in "order" out of chaos. In this context, self-organization is closely related to phase transitions as observed in the laser or the Bénard instability.

Even though the Langevin and Fokker–Planck equation are given as an example in Haken (1977) the chance to observe phase transitions in such systems seems to be very small. Consequentially, Scalapino et al. (1972) report for such systems: "[They] are nearly ordered but do not undergo sharp phase transitions." Furthermore, for a phase transition to occur high densities of the particles are necessary to establish a global coupling. However, in the swarm robotic scenarios investigated in this work such densities do not occur. Therefore, the theory of synergetics seems to be applicable to swarm robotics only as an abstract principle and the provided methods might prove to be inapplicable.

2.6 Philosophical Issues

Salomon saith:
There is no new thing upon the earth.
So that as Platon had an imagination
that all knowledge was but remembrance;
so Salomon gives his sentence
that all novelty is but oblivion.

Francis Bacon in "Essays LVIII"

2.6.1 Emergence and Novelty

It might be true that the concept of emergence is not applicable to solve complex design problems efficiently. Although there are (mainly natural) systems consisting of simple parts with simple behavior solving complex problems — this is not to be questioned — emergence might still be undesignable. Natural self-organizing systems are a result of evolution. Hence, a process without an explicit "outside purpose". Emergent systems just appear. Even artificial systems with emergent behavior are sometimes discovered by accident which is, for example, true for Chris Langton's ant. If emergence is really unpredictable in a very strict sense then there is no hope for designing artificial emergent systems efficiently. However, things do not look that bad presently since there are first promising projects showing the successful design of emergence. Already the open question of predictability of emergence should be motivation enough for projects like the present one.

Is the first phase transition (following Johnson (2001) given in the introduction) that ever happened in the universe from liquid water to steam really a novelty in principle or is everything defined microscopically?

The Origin of the Concept of Emergence – Metaphysics

The main contribution to the early development of emergence theory is usually accredited to John Stuart Mill (Stephan, 1999; Mill, 1843). He wrote (citing from Stephan, 1999):

> All organised bodies are composed of parts, similar to those composing inorganic nature, and which have even themselves existed in an inorganic state; but the phenomena of life, which result from the juxtaposition of those parts in a certain manner, bear no analogy to any of the effects which would be produced by the action of the component substances considered as mere physical agents. To whatever degree we might imagine our knowledge of the properties of the several ingredients of a living body to be extended and perfected, it is certain that no mere summing up of the separate actions of those elements will ever amount to the action of the living body itself.

The idea of emergent effects, however, was originally described by his student George Henry Lewes (Stephan, 1999; Lewes, 1875). The purpose of this theory is the "attempt to answer the metaphysical question of how to describe the relationship between mental and physical states" (Schumacher, 2002) or short: the attempt to resolve the mind–body dichotomy. The metaphysical origin might cause an indisposition for the swarm intelligence engineer but one might feel better reading that emergence theory is fundamentally naturalistic in the sense of a non-reductive physicalism, that is the assumption that natural systems are composed of physical components. The original idea of emergence was a "concept of genuinely new kinds of properties produced by nature that cannot be reduced" (Schumacher, 2002).

In the following we will not discuss individual definitions of emergence as it would be either too selective or a voluminous work due to the high number of such. Fortunately, most of the definitions boil down to a few fundamental concepts which we will discuss. The essential concepts in the definitions of emergence are either irreducible or reduceable properties, novelty, and downward causation.

Irreducible Properties

The concept of irreducible properties is controversial but many definitions of emergence, for example, the early ones, make use of it. An example is this statement by Anderson (1972):

> The ability to reduce everything to simple fundamental laws does not imply the ability to start from those laws and reconstruct the universe. The constructionist hypothesis breaks down when confronted with the twin difficulties of scale and complexity. At each level of complexity entirely new properties appear. Psychology is not applied biology, nor is biology applied chemistry. We can now see that the whole becomes not merely more, but very different from the sum of its parts.

Irreducible properties are observed or just existent on the macro-level but can neither be described nor explained by the sum of its parts. This is the principle, many have at first in mind, when thinking of emergence without any bias. However, there is a variety of problems one has to face, if this is the definition of your choice.

At first, one has to deny that there will ever be a theory proving the considered micro-macro relationship. However, we just might not know yet the needed theory, as written by Broad (1925): "Within the physical realism it always remains logically possible that the appearance of emergent laws is due to our imperfect knowledge of microscopic structure or to our mathematical incompetence." It is an open question whether it is possible to prove that a certain property cannot be explained by any theory. As we know that non-existence proves are already within mathematically well-defined frameworks hard to do, it seems to be very implausible that this might be possible for the fuzzy domain of general micro-macro relationships. This leads to a theory-relative concept of emergence. Then a property is emergent only relative to a set of theories. However, this could prove to be a quite dynamic definition with properties receiving and abandoning the label of emergence with the ongoing of science. The sense of such a definition is questionable.

Often the concept of irreducible properties includes unpredictability. Then we face the same problems as above. There might always be an unknown theory delivering this prediction. Sometimes this concept is restricted to unpredictability before the first instantiation. Such an instantiation could, for example, be the first phase transition of some material. The determination of the governing "order parameter" of a phase transition is currently believed to be unpredictable in principle (cf. Section 2.5.2 on page 19) as suggested by the following statement of Prigogine (1997):

> As long as we consider merely a few particles, we cannot say if they form a liquid or gas. States of matter as well as phase transitions are ultimately defined by the thermodynamic limit. The existence of phase transitions shows that we have to be careful when we adopt a reductionist attitude. Phase transitions correspond to emerging properties. They are meaningful only at the level of populations, and not of single particles.

However, there seems to be no way of proving that. The problem of classifying something as "unpredictable" persists as there will be theories, for example, as in theoretical physics, predicting phenomena no human being has observed before.

In some definitions the quite limiting reduction is demanded that a property is emergent, if the isolated components do not have this property, as for example by Kubík (2003):

> Basic emergence then refers to a property of the system that can be produced by interactions of its agents (components) with each other and with the environment and cannot be produced by summing behaviors of individual agents in the environment.

However, then almost everything would be emergent, for example, almost every man-made machine. Thus, it seems to be more useful to allow that less complex systems do not show this property. However, then, on the one hand, we need to find a reasonable definition for complexity and, on the other hand, the concept of the transition leading from too simple systems to complex systems showing the considered property is unclear. For example, consider the termites building clusters of sticks (Resnick, 1994) or the ants building clusters of corpses (Theraulaz et al., 2002). In this scenario a single agent would produce the same outcome and have the same functionality. Therefore, this scenario could not be emergent. Such a natural system that exhibits a non-trivial rapid qualitative change in its behavior and complexity with increasing agent number seems not to be known.

If the emergent property on the macro-level is causing effects to the entities on the micro-level, this is called downward causation — and there is a problem again. Bedau (1997) summarizes his view on "strong emergence", which uses the concept of irreducable properties with downward causation, as follows:

> Although strong emergence is logically possible, it is uncomfortably like magic. How does an irreducible but supervenient downward causal power arise, since by definition it cannot be due to the aggregation of the micro-level potentialities? Such causal powers would be quite unlike anything within our scientific ken. This not only indicates how they will discomfort reasonable forms of materialism. Their mysteriousness will only heighten the traditional worry that emergence entails illegitimately getting something from nothing.

Reducible Properties

Different from the above section is Bedau's concept of "weak emergence": The emergent property is reducible to the components of the system (Bedau, 1997). With this statement alone, everything would be emergent again. But he furthermore assumes that the emergent property can be predicted only by simulation or by modeling every interaction on the microscopic level. However, as discussed above, it is an open question, whether it can be proven or not that something can only be predicted by simulation. Again there would be the problem of dynamic emergence labeling depending on the ongoing of science.

A back door could be to define it as "can only be predicted by simulation at the current state of research". This would be analogical to defining the set of tasks for which during the execution intelligence is required by "everything at which man is better then the computer" — a set that might become empty step-by-step. Obviously this is unsatisfying as well.

Besides this discussion on predictability the approach of reducible properties seems to be more promising as it corresponds better with our today's understanding of the world — especially as engineers. This is well stated by Weinberg (1995):

> But there are no principles of chemistry on their own, without needing to be explained reductively from the properties of electrons and atomic nuclei, and in the same way there are no principles of psychology that are free-standing, in the sense that they do not need ultimately to be understood through the study of the human brain, which in turn must ultimately be understood on the basis on physics and chemistry.

Compare this quote to the quotation of Anderson above in Section 2.6.1 on page 21. Even if we accept this minimalistic definition of emergence and ignore the problem of predictability for a moment we still run into a problem: If the emergent property is reducible what is actually the meaning of emergence? This meaning exists only outside the system as an abstraction or a concept helping to understand what is going on.

Thus, one has to choose: accept the downward causation, that is, deny causal closure of physical reality, or accept the epiphenomenal view, that is, the emergent phenomenon has a cause but no effect. Unfortunately, both options are unacceptable. We do definitely not want to deny the physical reality and, on the other hand, epiphenomenal emergence seems to be very harmless and of questionable relevance. If we do not want to drop the theory of emergence we need to have a look on alternative approaches not using reducibility or irreducibility. This is discussed in the following.

Novelty

Emergence always requires a concept of novelty because the foundation is the observation of a new quality on higher levels. Anderson (1972) writes: "At each level of complexity entirely new properties appear." The observation of such new behaviors or patterns requires obviously an observer making emergence subjective, that is, it is relative to an observer. However, this is an objectionable approach, following Crutchfield (1994b): "Emergence is meaningless unless it is defined within the context of processes themselves." Crutchfield defines the concept of "intrinsic emergence" to overcome the requirement of an observer (Crutchfield, 1994a,b). By intrinsic emergence the novelty is defined within the underlying system. An external referent is not needed. The novelty is defined as an emergent global information processing architecture within the system that was not available before. Thus, the system itself capitalizes on the emergent property. However, the question, which Crutchfield chose as the title for his paper: "Is anything ever new?" (Crutchfield, 1994b), persists, although he answers it positively and so does also Prigogine (1997): "Nature is indeed related to the creation of unpredictable novelty, where the possible is richer than the real."

The question whether a system's macroscopic state is really new or whether it is defined on the micro-level already before the instantiation stays unanswered. As discussed by Stepney et al. (2006): How could anything new actually emerge from something that is already existent? However, Bickhard and Campbell (2000) argue that this question is ill-posed.

2.6.2 Multi-particle Systems and Computation

Artificial computational devices, such as sensor networks or robot swarms, that are deployed in the physical world, can be interpreted as a special computational device that is merged with its environment due to sensory input that is fed into an information process. The importance of embodiment and of the resulting concept of "embodied computation" is often emphasized (Stepney, 2007; Hamann and Wörn, 2007c) leading to an interpretation of swarm dynamics as computation. This interpretation is in correspondance to regard thinking as the evolutionary internalization of movement (Llinás, 2002; Churchland, 2007). There are different theories treating this concept ranging from hardware-oriented (e.g., world-embedded computation) to more abstract, software-oriented approaches (e.g., natural computation).

World-Embedded Computation

The idea of world-embedded computation by Payton et al. (2001) is to embed processing elements (robots) into the environment. Global properties of the environment and the system are computed collectively, robustly and distributedly by the robot swarm as a whole. Following Payton et al. (2001) the main benefits in comparison to traditional central approaches are adaptability in rapdily changing environments because sensing and computation are not seperated and performed in one step. The deployed system merges with the environment by incroporating global features of the environment in its computations. For example, the communication loss using infrared devices is exploited to determine optimal traversal paths.

World-embedded computation is similar to the concept of "embodied systems" as proposed by Stepney (2007). A system that senses its environment, that manipulates the environment, and responds to changes in the environment is seen as a complex dynamical stigmergic feedback process (c.f. Section 2.1.7), that is, a system that is embodied in its environment. The motivation of this concept is that a transfer of some computational burden to the complex environment might be possible. Essentially, the behavior of natural swarms utilizing stigmergy can be interpreted as sharing the computation cost between the swarm and the environment. This devision of costs is propagated as a design principle for artificial stigmergic feedback processes by Stepney.

Natural Computation

In the field of "natural computation" natural complex phenomena are viewed as computational processes. Following Castro (2007) natural computation is research on one of the following subjects or a combination of several:

1. nature as a source of inspiration for the development of computational tools,
2. computers as means of synthesizing natural phenomena,
3. the use of natural materials to compute.

Castro (2007) motivates research in this field:

> Natural computing has taught us to think "naturally" about computation and also to think computationally about nature. All natural computing approaches yield novel and exciting capabilities for science as a whole, mainly computer science, engineering, philosophy, and the biosciences. Scientists and engineers have been gifted with a very rich paradigm for exploration; the combination of nature and computing. In a broad sense, natural computing has taught us that any (model of a) natural phenomenon may be used as a basis for the development of novel algorithmic tools for problem solving; a vast array of natural phenomena can be simulated/emulated in computers; and natural materials can be used for computation. The fruits of these explorations are continuously becoming new technological solutions and explanations to old and recent problems, and the full potential is far from being reached.

The use of natural phenomena as a basis for novel algorithmic tools is, for example, displayed by swarm intelligence. If we generalize this statement from mere algorithmic tools to any kind of computational and/or embodied problem solvers, we can view swarm robotic systems as novel computational tools to solve complex problems.

2.7 Summary

In this chapter we have defined the important concepts of agents, agent–agent interactions, macro- and micro-level, phenomenological models, self-organization, and emergence. The influences to swarm robotics from both fields, hardware and software, have been shortly summarized. The main influences related to hardware are micro-robotics and sensor/actuator networks. The main influences related to software are distributed artificial intelligence, multi-agent systems, artificial life, and amorphous computing. We have summarized the two mature theories of self-organization (in physics), namely the theory of dissipative structures and synergetics. Finally, we have discussed the philosophical issues of emergence, novelty, and the interpretation of multi-particle systems as computational systems.

Chapter 3
State-of-the-Art in Modeling and Design of Swarms

To understand emergence,
we must understand the way
in which models in science and elsewhere
allow us to transcend the knowledge
that went into their construction.

John H. Holland in "Emergence - From Chaos to Order"

We summarize the state-of-the-art in modeling swarms in several relevant fields and discuss the challenge of designing emergence.

In the first two parts of this chapter, a variety of models originating in different fields is presented. All these models might potentially be applied to swarms of robots although they were originally developed to describe systems or swarms of molecules, insects, mammals, people, or only small groups of robots.

In the third part, the current state-of-the-art in designing emergent behavior, that is, the design of control algorithms for individuals of decentral systems operating on the basis of local information only, is presented.

3.1 Modeling Swarm Behavior and Collective Behavior

3.1.1 Agent-Based Modeling

Models that explicitly implement each individual agent of a system, for example, by a finite state machine and agent-centered information (e.g., position, speed) are called agent-based models. Research in agent-based modeling dates back to the self-reproducing automata of Neumann (1966). Due to limited computational resources it was unpopular until its breakthrough in form of the new field of artificial life

H. Hamann: Space-Time Cont. Models of Swarm Rob. Sys., COSMOS 9, pp. 27–39.
springerlink.com

founded by Langton (1989b) (see Section 2.4.3 on page 17). Since then, agent-based modeling has been applied to many fields ranging from logistics to social networks.

3.1.2 Classical Approach – Control Theory

There are many studies, which apply control theory to model and control robotic swarms, for example see Gazi and Passino (2003) or Feddema et al. (2002). One perspective indicates that the preconditions for these approaches are very limited. Full synchronization, global communication or communication to a fixed group of neighbors and deterministic behavior are often assumed. However, this approach allows extensive stability analysis and certainties. But the nondeterministic component, strictly local communication, and the possibility of coherence breakdowns are very relevant in swarm robotics and are not addressed in these control theoretic approaches.

3.1.3 Swarm Robotics

In swarm robotics there is only a small variety in mature state-of-the-art models ready to support the algorithm design phase despite great endeavors lasting for almost a decade. Martinoli (1999) presents a modeling technique based on rate equations which was successfully applied to several scenarios. These rates quantify fractions of the swarm being in the same state. The temporal evolution of such a rate is, in most cases, easily derived and typically based on geometric considerations.

Many works can be found that followed this approach by Martinoli himself but also by Lerman et al. (Martinoli et al., 2004; Lerman and Galstyan, 2002; Lerman, 2004; Lerman et al., 2005, 2006). A methodologically very sophisticated approach is reported in (Correll and Martinoli, 2006; Correll, 2007) where parameters of the rate equations are determined by the use of system identification techniques.

A drawback of these rate equation approaches is the limited representation of space – either homogeneous space is assumed or a rough discretization by states is done. Galstyan et al. (2005) present a generalization of the beforehand applied type of Master Equation (i.e., the foundation of the rate-equation approaches) for robot densities in continuous space. Similar to the biological models (see Section 3.1.6 on page 32), this work is also a phenomenological macroscopic approach without the establishment of a connection to the microscopic level. Furthermore, no comparison to simulations is presented.

Many swarm robotic scenarios are based on heterogeneous space, for example, aggregation, exploration of non-periodic environments, transport, and collective movement. In these applications, an explicit representation of space in the model is helpful, if not necessary. Such models can be found in physics. They describe the motion of many entities or particles in continuous space. Typically, there are microscopic as well as macroscopic models available for the same phenomenon, for example, kinetic theory and thermodynamics. At least for extensive robotic swarms it is adequate to model a robot as a particle, that is, essentially reducing the geometry of a robot to a cross section. The equations provided by physics are often not analytically solvable. Therefore, methods of numerical analysis have to be applied. In turn, this leads back to discrete representations due to the limitations of our computers. This discretization can, however, be refined arbitrarily when provided with enough resources (computational power and memory). Finally, the equations effective in each point of the discretized space might be understood as rate equations of swarm fractions entering and leaving the associated area.

In addition to the rate equation model, several other preliminary or rather specialized models were proposed: Winfield et al. (2005) present a formal method using temporal logic to specify and possibly prove emergent swarm behavior. A related approach is reported by Kornienko et al. (2005b) that is also formal. They derive local rules such that the robots form predefined shapes. Soysal and Şahin (2007) analytically derive a model for a special case of aggregation behavior by applying combinatorics and linear algebra. Bjerknes et al. (2007) present a model for a special case of the collective motion algorithm reported by Nembrini et al. (2002). This algorithm is discussed in this work as well (see Section 5.3 on page 100). Compared to its inherent expressiveness, their model is surprisingly simple but mainly based on directly measured agent-agent interaction frequencies and durations in the complete swarm. Furthermore, see the work of Milutinovic and Lima (2007) for an approach focusing on stochastic models inspired by immune systems.

3.1.4 Physics

In physics very sophisticated models of multi-particle systems have been developed. One prominent achievement is the sensitive distinction between models of trajectories and models of ensembles, which is discussed in the following. Thereafter, we give an overview of physical models that were applied to systems of autonomous agents.

The Modeling Question: Trajectories or Ensembles?

The more intuitive approach of modeling multi-particle systems might be models based on trajectories. For simplicity, one starts with the easiest case: one particle moving deterministically. This single particle can easily be modeled. Adding a second particle introduces a new quality: particle interactions. These two particles move around just as a single one until they interact, that is, they collide. We assume that for a collision there is a deterministic rule to determine the resulting velocities. After the collision the particles are correlated, that is, for all time their velocities and positions will be partially determined by this collision. We call this a binary correlation. By introducing more and more particles to the system, particles will be correlated to three, four, and more particles, that is, they are ternary, quaternary, …, N-ary correlated. Collided particles will be separated over time and form long-range correlations due to their history. Such long-range correlations, however, might become influential to the system's behavior. The particles form a history of correlations, which might lead to global patterns and is interpreted as a collective effect. The underlying principles of this global effect might be masked by a model based on trajectories because it is concealed by microscopic details.

Therefore, a model of ensembles, that is, treating an average over many systems states at once (e.g., probability densities), might be beneficial. By the group around Prigogine the ensemble approach is even propagated to be obligatory in the thermodynamic limit (infinite number of particles), that is, the reality is not presentable using trajectories (Prigogine, 1997; Bishop, 2004). However, in this work we consider finite agent numbers only. Still the ensembles approach is an attractive option because the robot's motion is noisy and we are interested in the behavior of the swarm averaged over all initial configurations.

Models

In our model, we decided to follow the modeling approaches originating in physics. Based on Brownian motion Schweitzer (2003) builds up the theory of Brownian agents, that is, self-propelled active particles. In the scenarios investigated by Schweitzer the Brownian agents typically interact only indirectly via a potential field and not directly, for example, not through physical interactions or via direct communication (Schweitzer et al., 1997; Schweitzer, 2002, 2003). Thus, the Langevin and the Fokker–Planck equation can be directly applied to Brownian agents without shortcomings. Furthermore, most of Schweitzer's research is based on microscopic models that he favors over continuous approaches mainly due to their smaller computational overhead. However, if one wants to take physical interactions into account this computational advantage can be at least equalized.

Similar approaches are: Helbing et al. (1997) and Vicsek et al. (1995) as well as a related recent work by Hogg (2006). In all three papers, the authors discuss the control of microscopic robots by Brownian motion, gradient ascent, and cooperation. While Hogg (2006) presents a phenomenologically motivated model, Vicsek et al. (1995) intensively use, similar to Schweitzer, microscopic simulations instead of macroscopic models. Helbing et al. (1997) study macroscopic models for a special scenario (formation of trail systems) derived out of a microscopic description. However, these are sophisticated derivations adapted to the considered special scenario and a generalization of this approach seems to be improbable. In addition, all of these approaches except Hogg (2006) lack direct agent–agent communication (e.g., via radio using a type of non-trivial protocol).

3.1.5 Chemistry

Collision Theory

Collision theory is a qualitative description of chemical reactions and offers the possibility to estimate the expected number of collisions for moving solid particles

(Trautz, 1916; Hänel, 2004). First, consider the situation of a particle f moving at velocity v and passing an area occupied by unmoved particles (see Figure 3.1). Particle f covers a distance of $v\Delta t$ in a time of Δt and collides with an unmoved particle a if $\text{dist}(f,a) \leq \sigma$, for the reaction cross section of σ (a diameter). For now we set $\Delta t = 1$. Thus, the relevant area covered by a moving particle is

$$C_{m,u} = 2\sigma v, \tag{3.1}$$

which is the significant area to calculate the number of moving-to-unmoved collisions. By multiplying $C_{m,u}$ with the number of unmoved particles in a considered area divided by the size of this area, that is, a density of unmoved particles, we get the number of expected collisions for a single free robot.

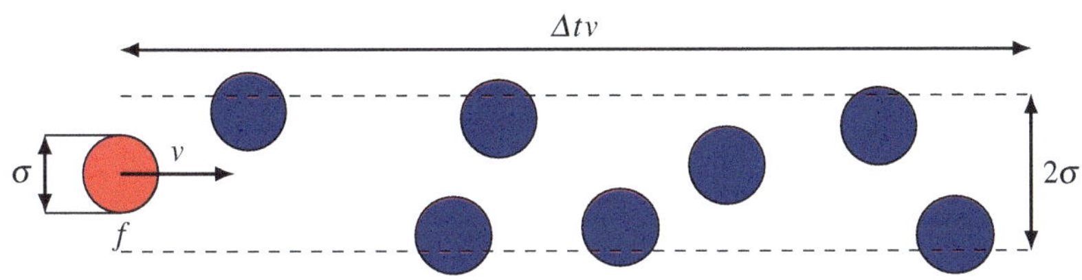

Fig. 3.1 Concept for computing parameters based on collision theory, reproduced after Hänel (2004)

Second, consider the situation of a moving particle f passing an area populated by other moving particles, all moving at the same velocity v. Now, the situation is different from that above as we have to take the relative speed into account. For example, two particles traveling in the same direction keep their distance and do not collide, while others heading for confrontation collide within a short time even for big initial distances. This is accounted for by summing the relative speeds over all uniformly distributed velocity vector configurations resulting in $\overline{v} = \frac{4v}{\pi}$. By using this mean relative speed $\overline{v}$ instead of v we get the equivalent of Equation 3.1 for moving-to-moving collisions:

$$C_{m,m} = \sigma \frac{4v}{\pi}. \tag{3.2}$$

By multiplying with the density of moving particles we get the expected number of colliding particles which is twice the number of collisions.

Reaction–Diffusion Systems

Reaction–diffusion systems are spatially extended models of constituents that localy interact through chemical reactions and that are transported in space by diffusion processes (Kuramoto, 1984). Two-component reaction–diffusion equations (systems of two constituents) can show bifurcations and pattern formation (Turing,

1952). Reaction–diffusion systems are mathematically represented by partial differential equations of the form

$$\frac{\partial \mathbf{x}}{\partial t} = \mathbf{R}(\mathbf{x}) + \mathbf{D}\nabla^2 \mathbf{x}, \tag{3.3}$$

for a vector $\mathbf{x}$ of the substances' concentrations, a function $\mathbf{R}(\mathbf{x})$ describing the local chemical reactions, and the diffusion coefficients in form of a diagonal matrix $\mathbf{D}$.

Although reaction–diffusion systems were originally applied in chemistry, they are applied to many fields today (Murray, 2003; Eigen, 1979). An example is Fisher's equation

$$\frac{\partial x}{\partial t} = x(1-x) + \nabla^2 x, \tag{3.4}$$

which is a one-dimensional representative of a reaction–diffusion system with $R(x) = x(1-x)$ and $D = 1$. The original application of this equation was population genetics (Fisher, 1930).

3.1.6 Biology

While in physics sophisticated and mature models exist, that describe the considered systems extensively, in biology models seem to be rather incomplete and preliminary. This is most likely due to the high complexity of biological systems that build up on physics. Theories exist for the phenomenon of self-organization in physics (see Section 2.5 on page 18). In biology such extensive approaches are not within grasp. Again self-organization seems to be a more complex phenomenon in biology compared to physics (Leiber, 2000).

Still it may appear obvious to use agent models from biology in swarm robotics; however, models of robotic swarms differ from biological models in their synthesis. In the case of swarm robotics, an extensive knowledge of the agents' interior exists because they have been constructed. In addition, robots typically have the ability to communicate exact data via electromagnetic radiation, which distinguishes them from a majority of biological models. Nevertheless, biological models might become more relevant, if minimalist hardware is applied, and an extensive progress has been achieved in the field of mathematical biology (Okubo, 1986; Grünbaum and Okubo, 1994; Okubo and Levin, 2001; Edelstein-Keshet, 2006; Mogilner and Edelstein-Keshet, 1999). Okubo (1986) provides a good review of a variety of swarm modeling concepts. Typically, the macroscopic approaches are phenomenological, that is, a macroscopic description is derived based on observations of the swarm behavior. However, "many diverse rules can lead to aggregates that look and behave quite similarly" as discussed by Edelstein-Keshet (2006).

The microscopic approaches are based on simulations using models of individuals. These models are verified by observing the resulting macroscopic behavior (Deneubourg et al., 1990; Bonabeau et al., 1998; Theraulaz et al., 2003). This approach is in contrast to what we pursue in this work. We design the local rules ourselves and derive a macroscopic model with a strong connection to the microscopic level.

A new approach is to use robot swarms as embodied models of biological systems (Hamann et al., 2008; Schmickl et al., 2007c,a; Schmickl and Crailsheim, 2006). The work with hardware exposes difficulties unobserved and uncared in the simulation, such as limited accuracy of sensors or actuators. Thus, robot swarms serve as a stricter validation than the mere comparison of simulation results to the observed behavior in the natural system. Hence, the process of investigation is roughly: observation of the natural system, modeling, implementation in robotic hardware, observation of this artificial system, and correction of the model. This approach might lead to a better understanding of biological systems in the future.

3.1.7 Micro, Macro, and Stochastic Modeling – Road Traffic

Modeling of road traffic is an intensively studied subject which is indicated by a vast number of publications and different modeling techniques. The economic idea is to maximize the throughput while minimizing the cost for roads and highways. Obviously there is a tradeoff. For that reason a lot of time and money was invested in models that predict the traffic in road networks.

The multitude of road traffic models can be classified by three categories:

- Microscopic: Every single road user is represented in the model, the individual behavior can be observed in the model.
- Macroscopic: A continuous model approach that represents the traffic as a whole by a fluid in "road canals".
- Stochastic: The focus is on crossing points, the behavior of individual road users between these points is of less importance. Each crossing point is modeled by a waiting queue.

The relevance of single road users is indisputable. For example, a single agricultural utility vehicle might cause a traffic jam. Early models of road traffic where focused on the macroscopic approach. However, in the early 1990s researchers turned to *microscopic models* (Nagel and Schreckenberg, 1992) because they promised to be a feasible and fast simulation tool of large networks (an argument which is partially obsolete today). These microscopic models are based on cellular automata that discretize streets or lanes by sites of, for example, 7.5m length in the case of a highway. Each of these sites holds maximally one car. Faster traffic from behind will be slowed down and has to queue. Such models are rich enough to explain, for example, the problem of "traffic jams out of nothing".

The idea of *macroscopic models* of traffic is based on the observation that disturbances (e.g., the tail end of the traffic jam at a traffic light) spread wavelike. The first model of that kind was published by Lighthill and Whitham (1955). Typically such a model is space- and time-continuous and the three most important variables are: traffic speed $v(x,t)$ (position x, time t), car density $\rho(x,t)$, and traffic flow $f(x,t)$. Obviously, there will be a maximal speed $v_{\max}$ and a maximal density $\rho_{\max}$. One of the most prominent results is the fundamental diagram of traffic flow that describes the relation between traffic flow and density

$$f(\rho) = v_{\max}\left(1 - \frac{\rho}{\rho_{\max}}\right)\rho. \tag{3.5}$$

This function is shown in Fig. 3.2. The most desirable density is ρ_{opt} which results in the maximal traffic flow $f_{\max}$ (i.e., the most efficient utilization of the road). Furthermore, it is interesting that for $\rho < \rho_{\text{opt}}$ there are always two density options of equal flow (e.g., $f(\rho_1) = f(\rho_2) = f_{12}$) that correspond to two different speed limits. Hence, the traffic manager has two options while for the road user the higher speed limit is always more desireable.

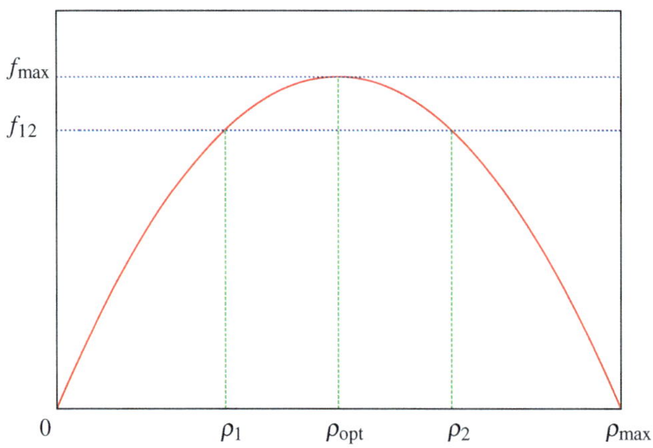

Fig. 3.2 Fundamental diagram of traffic flow

In *stochastic models* of traffic the focus is on crossings and the connecting roads between crossings are abstracted by stochastic processes. Only the arrival times of cars at a crossing point are considered. The cars are inserted in a waiting queue and leave the crossing after they have been "processed". Hence, stochastic models of road traffic are similar to models of data network traffic or, for example, printer queues.

The most interesting approach of traffic models, that is related to the work at hand, is the macroscopic modeling approach, that is the abstraction of individual cars as a fluid with a certain density. However, the traffic model approach is limited by the assumption that cars drive only on a given street network. These networks a constructed by one-dimensional building blocks.

3.2 Design of Emergent Behavior

3.2.1 Global-to-Local Programming – Engineering of Emergence

The challenge in engineering swarm robotic systems is the problem of engineering algorithms for a distributed system that processes only local information but achieves a globally defined task. Thus, the challenge is to engineer system showing emergent properties. We call this challenge "global-to-local programming" as it is called by Yamins (Yamins, 2005, 2007; Yamins and Nagpal, 2008). We regard the problem of determining the "global-to-local connection" or the "local-to-global connection" as the same problem and, thus, as interchangeable terms. To find local rules for a preassigned task or to determine the global behavior based on the local rules are not really two approaches to a problem but essentially just one approach because "global-to-local" cannot exist without "local-to-global" and vice versa. To determine the class of patterns a local rule can produce or to determine the local rule for a given pattern is the same. In both ways either the given or the designed rule needs to be tested to check the resulting global behavior and, in fact, it will typically not just be one check but rather an iterative process alternating between the design or investigation of the local rules and the investigation of the global behavior.

The challenge of engineering emergence is a problem in general and partially due to its interdisciplinary form, as diagnosed by Haken (1999): "[...] despite a lot of knowledge about complex systems the application of this knowledge to the engineering domain remains difficult. Efforts are scattered over many scientific and engineering disciplines." It begins with the trivial problem of different vocabulary in the involved fields of research. The incapability of finding a well definition of the actual challenge in designing emergence is part of the problem as well and the development of general approaches or a general theory seem to be very difficult (Stepney et al., 2006; Abbott, 2006). Parunak et al. (2005) give a good summary of the present difficulty:

> Many multi-agent systems seek to reconcile two apparently inconsistent constraints. The systems overall objective is defined at a global level. However, the agents have only local information available to them in selecting their actions. Such systems are presently more art than science. They often exhibit regularities (such as exponential convergence) that we do not understand, and we do not know how to improve their functioning in a disciplined manner.

Whether engineering of emergence is possible at all is an open question. Stepney et al. (2006) express their hopes: "[The] philosophical discussion gives us the hope, at least, that engineering emergence is not an impossible dream." Abbott (2006) shows also optimism in this issue, although he excludes fully automatic approaches because he detects the need for a creative step.

3.2.2 Antagonism of Concepts

There might be an antagonism in the term "engineering of emergence". Systems showing emergent properties are mostly considered to be self-organized as well.

Thus, engineering of emergence means mostly also engineering of self-organization. However, a definition of self-organization states that it is organizing itself without any exterior influence. We want a system to be self-organizing but the task it should solve is defined by ourselves — from the outside.

Besides this rather obvious dichotomy there is another problem, too. Emergent properties are often defined to be unpredictable (Bonabeau, 2002; Deguet et al., 2006). However, if Abbott is right and a creative (or constructive) step is necessary in the engineering of emergence then a feasible iterative design principle cannot exist because the systems needs to be implemented to observe the resulting global behavior. The situation would be very different, if emergence is unpredictable but an arbitrary accurate approximation is possible. This would be sufficient to usefully apply any kind of constructive methods to engineer emergence.

3.2.3 *Programming "by Hand"*

In classical software engineering formal methods such as stepwise program refinement are applied that transform an abstract specification into a concrete executable program. Such methods cannot be applied to the swarm robotic domain because it is beliefed that local rules, which generate emergent behavior, cannot be derived formally as they need a constructive step (see Section 3.2.1 on the page before). Furthermore, predicting the resulting global behavior for a set of local rules is at least difficult, if not impossible.

The simplest approach to solve this problem is the direct approach without any tools. The local control algorithm is programmed just "by hand" and tested in a second step using the robotic hardware (see also Section 4.2.1 on page 57). Doing this directly on the hardware is very costly concerning time and wear. Therefore, typically a simulation is applied. The development of control algorithms supported by simulations is easier and faster. For example, the possibly laborious upload of the executable program is not required, debugging is much more convinient in a simulation framework, the implementation is not delayed by hardware failures, etc. However, the virtual domain (simulation) and the hardware domain (robots) should be ideally very similar to ensure that programs working in the simulation will work on the real robots as well. This demand necessitates a subtle robot model and possibly also the complete simulation of relevant physical forces (physics engine). However, an elaborated robot model and a physics engine introduce high computational complexity (e.g., raytracing for the simulation of sensors, iterative numerical methods to compute forces). A partial solution is to introduce several robot models of different abstraction level. Thus, one would test a program at first with the most abstract model. Then the program is (hopefully) stepwise adjusted to the different robot models in order to correct the probably different global behavior, see Figure 3.3 on the next page. Finally, the program is tested with the robot hardware and adjusted for the last time. This is the state-of-the-art software engineering of swarm robotics: an iterative or trial-and-error process based on a hierarchy of simulations and robotic hardware.

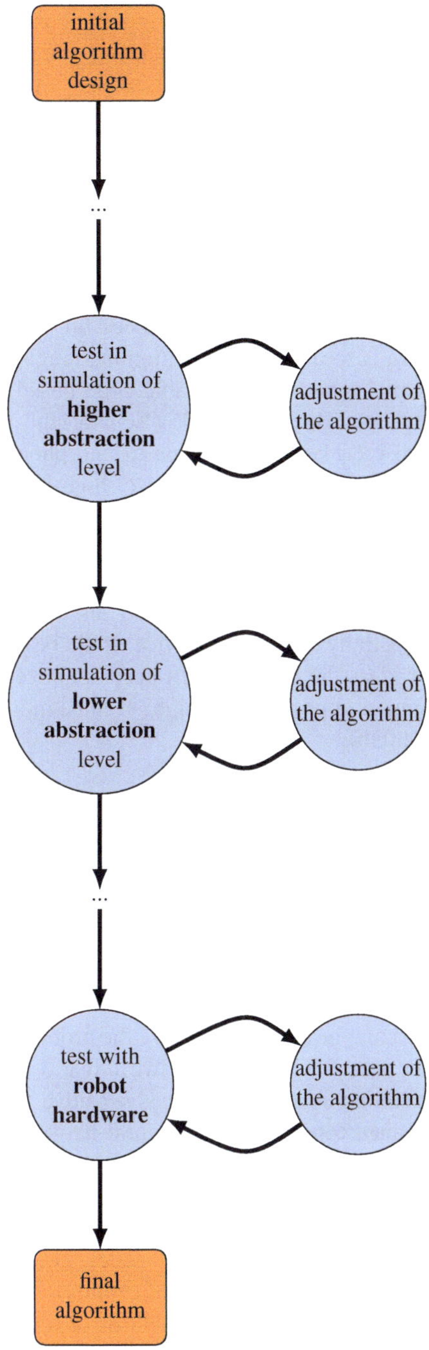

Fig. 3.3 Iterative algorithm design based on a hierarchy of simulations and robotic hardware

3.2.4 Automatic and Semi-automatic Methods

Besides the direct approach towards swarm robot programming there are several approaches generating executable code automatically or semi-automatically. In the following we give three distinguished examples.

Amorphous Computing

The amorphous computing approach (see Section 2.4.4 on page 17) to global-to-local programming is a formal approach to the design of emergence. Whether such an approach can be successful was discussed in Section 3.2.1 on page 35. The idea is to design a programming language ("Proto") and a compiler that automatically generates executable code based on a rather abstract (spatial) describtion. Beal and Bachrach (2006) "separate the engineering problem into three components: a discrete kernel to emulate an amorphous medium and distribute code, a compiler for Proto, and implementations of high-level coordination and homeostasis primitives, allowing simple and concise expression of programs controlling spatial behaviors."

However, the transition from an abstract description to a hand coded program is blurry. The design process is simplified by introducing the abstraction of a spatial medium that is programmed. Thus, the algorithm designer defines the algorithm as if a spatially continuous device would be programmed. However, within this construct of ideas the designer still needs to have a valid intuition of the global behavior. A similar approach that might even be more formal was proposed by Kornienko et al. (2005b).

Evolutionary Algorithms

"Evolutionary algorithms" is a collective term for metaheuristic optimization algorithms inspired by biological evolution, such as genetic algorithms, genetic programming, or evolutionary programming. In order to apply such methods to the algorithm design problem, it is transformed into an optimization problem.

Depending on the actual method this is done by defining a metric/fitness function and maybe also by reducing the problem to the problem of finding optimal weights for an artificial neural network that defines the robot behavior. The validity of the resulting swarm behavior is sensitive to the fitness function or the (possibly) predefined topology of the neural network. Furthermore, this approach is extremely intensive concerning computational complexity (up to several days of CPU-time for a simple behavior) and it might also be inapplicable to more complex behaviors due to an exponentially increase of necessary computational power (due to an exponential increase of the search space). The frequently employed counter-argument "but it worked for the biological evolution" is not applicable because evolutionary algorithms are extreme abstractions of the natural complexity. Even if an appropriate algorithm is found a problem might arise, citing Hillis (1999):

> One of the interesting things about the sorting programs that evolved in my experiment is that I do not understand how they work. I have carefully examined their instruction sequences but I do not understand them: I have no simpler explanation of how the programs work than the instruction sequences themselves. It may be that the programs are not understandable.

The value of a system that is not understandable is, unfortunately, questionable, especially, concerning warranties.

Nevertheless, evolutionary algorithms have been applied frequently to the swarm robotic domain and with success at least for the generation of simple behaviors (Trianni, 2008; Dorigo et al., 2004).

Machine Learning

Machine learning techniques can be interpreted as optimization algorithms, too. The provided methods include: policy iteration, value iteration, dynamic programming, Q-learning etc. They can be applied directly on the hardware (online) and/or in a training phase in a simulation beforehand. Using online learning means the robot swarm attains adaptation as a form of "lifelong learning". However, the learning techniques suffer, just as evolutionary algorithms, from the exponential increase of the search space with increasing behavior complexity. This is especially true for cooperative multi-agent learning (Panait and Luke, 2005). To the knowledge of the author machine learning techniques have not yet been applied to real robotic swarms. This might be due to the fact that the majority of the swarm robotics community is in favour of or at least influenced by minimalist swarm robotics (see Section 2.3.3 on page 14). There are works applying machine learning techniques to small groups of robots, see for example Matarić (1997).

3.3 Summary

We have presented the state-of-the-art in swarm models of related fields. This included the general concept of agent-based modeling, control theory, swarm robotics itself, physics, collision theory, and biology. In the second half of the chapter, we have discussed the challenge of designing emergence and have highlighted alternative approaches of constructing swarm control algorithms, namely programming "by hand", evolutionary approaches, and learning.

Chapter 4
A Framework of Models for Swarm Robotic Systems

Thou turn thy mind the more unto these bodies
Which here are witnessed tumbling in the light:
Namely, because such tumblings are a sign
That motions also of the primal stuff
Secret and viewless lurk beneath, behind.
For thou wilt mark here many a speck, impelled
By viewless blows, to change its little course,
And beaten backwards to return again,
Hither and thither in all directions round.

Titus Lucretius Carus in "On the Nature of Things"

We introduce the two basic equations of the model, present the model framework, give clues how it is applied, discuss a computational approach to swarms, and give an overview over the limitations and the benefits of our model.

In this chapter the two basic equations of the model are introduced and the model framework itself is motivated and presented. Parts of this chapter have been previously published in Hamann and Wörn (2008b) and Hamann and Wörn (2007b). In Section 4.7 on page 68 the limitations and possible benefits of the application of the presented model as a model of computation are discussed.

4.1 Modeling Multi-particle Systems – Example of Brownian Motion

In this section, a method of modeling multi-particle systems based on physics is presented. This modeling technique applied to multi-agent systems is relevant to today's research in this field, although the roots of these methods date about 100 years

H. Hamann: Space-Time Cont. Models of Swarm Rob. Sys., COSMOS 9, pp. 41–76.
springerlink.com

back. In particular, the following methods form the basis of the model proposed in this work.

4.1.1 Introduction – History of Brownian Motion

Brownian motion is named in the honor of Robert Brown, although the phenomenon was vaguely reported previously, for example, by Lucretius in "On the Nature of Things". Brown (1828) reported his observations of particles within the vacuoles of pollen grains under the microscope. He observed the now well-known random motion of a particle bigger by one order of magnitude compared to the size of the molecules in which it floats. The phenomenon of Brownian motion was mathematically solved by Einstein (1905) in one of his Annus Mirabilis (Einstein et al., 1998) papers and independently by Smoluchowski (1906). Three years later, Langevin (1908) discovered a different and easier approach with the microscopic description of Brownian motion – citing from (Lemons and Gythiel, 1997, a commented translation of Langevin's paper):

> While Einstein, starting from reasonable hypotheses, derived and solved a partial differential equation [...] governing the time evolution of the probability density of a Brownian particle, Langevin applied Newton's second law to a representative Brownian particle. In this way, Langevin invented the $F = ma$ of stochastic physics now called the *Langevin equation*.

Fokker (1914) and Planck (1917) then analyzed the probability distribution of the velocity in the diffusion process by deriving a partial differential equation now called Fokker–Planck equation — a macroscopic description. However, this was only the first use of this equation which has an extensive variety of applications. This is also indicated by the independent rediscovery of this equation by Kolmogorov (1931). In his honor it is called the Kolmogorov forward equation describing time- and state-continuous Markov processes. Later, it was shown that the Fokker–Planck equation can also be applied to systems far from thermodynamic equilibrium. It is used to describe noise in quantum optics, for example, in modeling a laser (Risken, 1984) which can also be considered a self-organizing system as investigated in synergetics founded by Haken (1977). Schweitzer (2003) developed the concept of Brownian agents which are active particles with an internal energy depot and self-driven motion. For such reactive agents a generalized Langevin equation can be derived to describe the agents activities and offers possibilities of analysis including the formal derivation of a Fokker–Planck equation.

4.1.2 A Microscopic Model – The Langevin Equation

Despite its microscopic description, the Langevin equation is already an abstraction. Only the trajectory of a single particle is described. The other particles, in which it is floating, are not modeled explicitly but summarized in a stochastic process.

There is not "the Langevin equation" as "Langevin equation" is used as a label for a class of equations of similar form. However, there is one original Langevin equation as given by Langevin (1908) to describe the trajectory of a particle under consideration (for two dimensions: velocity $\mathbf{v} = (v_1, v_2)^T$ and position $\mathbf{R} = (R_1, R_2)^T$):

$$\mathbf{v} = \frac{d\mathbf{R}}{dt}, \qquad m\frac{d\mathbf{v}}{dt} = -\gamma\mathbf{v} + \Psi(t), \tag{4.1}$$

for mass m, friction constant γ (also called resistance coefficient), and a noise term $\Psi = (\Psi_1, \Psi_2)^T$ which describes the random forces acting on the particle and which is a stochastic process. Thus, the equation itself describes a stochastic process. The Langevin equation is a stochastic differential equation. The physicist Langevin was ahead of time compared to the mathematicians as the mathematical interpretation of such equations was unclear until a formalism was established in the 1940s by Itô (1951). For the case of Brownian motion Langevin was able to determine some evident properties of Ψ. As Ψ describes the complementary force that keeps the particle in motion against its friction, the strength of Ψ can be determined. Furthermore, the irregularity of this force is well known and, thus, we know that the average force becomes $\langle\Psi\rangle = 0$. Starting with these assumptions he obtained the same results as Einstein and Smoluchowski.

Now we extend this notation by a potential field $P(\mathbf{r},t)$ with $\mathbf{r} = (r_1, r_2)^T$ denoting space. We assume this potential (e.g., electric field, constant flow in a liquid, steep surface) causes a force proportional to its gradient in the particle. We get a Langevin equation describing Brownian motion with drift in a potential field by extending Equation 4.1 to

$$\mathbf{v} = \frac{d\mathbf{R}}{dt}, \qquad m\frac{d\mathbf{v}}{dt} = -\gamma\mathbf{v} + \left.\frac{\partial P(\mathbf{r},t)}{\partial \mathbf{r}}\right|_{\mathbf{R}} + \Psi(t). \tag{4.2}$$

On reasonable large timescales the decay time of the velocity of the Brownian particle (typically $1/\gamma = 10^{-7}s$) can be neglected. For times $t \gg 1/\gamma$ Equation 4.2 can be reduced to the Einstein–Smoluchowski limit (Hauge and Martin-Löf, 1973; Schweitzer and Schimansky-Geier, 1994; H'walisz et al., 1989):

$$\frac{d\mathbf{R}}{dt} = -1/\gamma \left.\frac{\partial P(\mathbf{r},t)}{\partial \mathbf{r}}\right|_{\mathbf{R}} + \mathbf{F}(t), \tag{4.3}$$

with a noise term $\mathbf{F}$, which is interpreted as a random dislocation.

Now we leave these purely physically motivated models and get to a more abstract model of Brownian motion with drift. In exchange, several statistical properties can be determined. We use the concept of a *Wiener process* W_t, that is a stochastic process obeying the following three conditions:

1. $W_0 = 0$,
2. W_t is *almost surely* continuous,
3. W_t has independent increments with distribution $W_t - W_s \sim \mathcal{N}(0, t-s)$.

In one spatial dimension we get as a model for Brownian motion with drift:

$$R_t = at + bW_t, \tag{4.4}$$

with drift a and volatility b defining the intensity of the stochastic motion. The drift term at describes a constant drift as it could result from a constant force pulling the particle. The following property follows directly

$$R_t - R_s \sim \mathcal{N}\left(a(t-s), b^2(t-s)\right). \tag{4.5}$$

A Wiener process shows self-similarity. For example, the properties of a Wiener process are invariant to the timescale (all processes $R_t = \frac{1}{\sqrt{\alpha}} W_{\alpha t}$, for $\alpha > 0$ are Wiener processes).

For simulating a Wiener process numerically we can use the fact that a specially defined random walk converges to a Wiener process. A random walk is defined by

$$R(t+\tau) = R(t) + \Phi(\tau), \tag{4.6}$$

for step size τ and a random variable Φ. We define $R(0) = 0$. Say $\Phi(\tau)$ is defined such that with probability $1/2$ it is $\sqrt{\tau}$ and with probability $1/2$ it is $-\sqrt{\tau}$. Now this random walk converges to a Wiener process for $\tau \to 0$ (Donsker's invariance principle). Thus, for a numerical simulation of a Wiener process we can use the following approximation

$$W_{n\Delta t} \approx \sqrt{\Delta t} \sum_{i=1}^{n} B_i, \tag{4.7}$$

for Bernoulli-distributed B_i. From Equation 4.7 follows the update rule

$$W_{n\Delta t} - W_{(n-1)\Delta t} \approx \sqrt{\Delta t} B_n, \tag{4.8}$$

Based on these considerations we can now define a concluding Langevin equation. Say we have just one dimension and the noise is a sequence of uncorrelated impulses of strength one. Thus, we define

$$F(t) = \sum_j \delta(t - t_j)(\pm 1)_j, \tag{4.9}$$

for the Dirac delta function δ and a sequence of impulses $(\pm 1)_0, (\pm 1)_1, (\pm 1)_2, \dots$ at times $t_0, t_1, \dots$ Thus, for an unspecified drift A and a diffusion coefficient B we get the Langevin equation

$$\frac{d\mathbf{R}}{dt} = A + BF(t). \tag{4.10}$$

This equation is numerically simulated by the stochastic process given in Equation 4.4 using the Wiener process, which is itself simulated by Equation 4.7. For examples of sampled trajectories see Figure 4.1 on the next page. The C-program, that was used, is given in Appendix A on page 127. It turns out that any distribution with mean zero and variance one can be used in the numerical

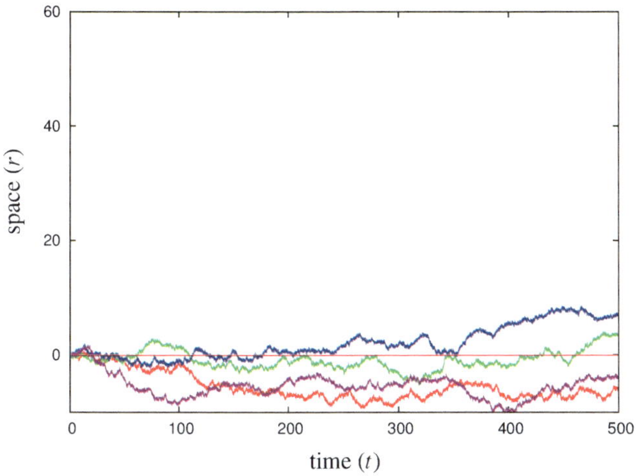

(a) Four trajectories with $A = 0$ (no drift).

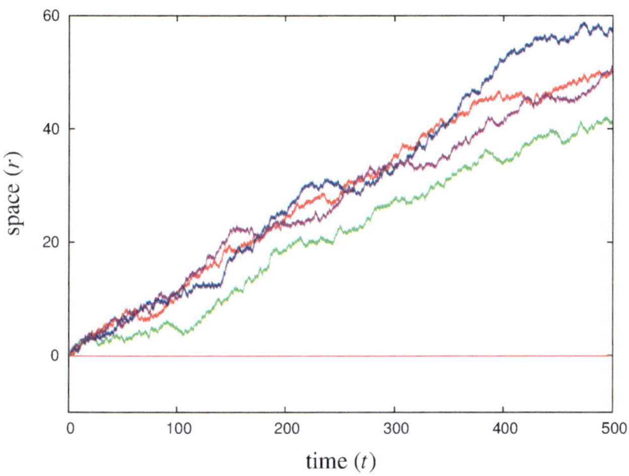

(b) Four trajectories with $A = 0.1$ (with drift).

Fig. 4.1 Sampled trajectories in 1-D without and with drift using Langevin equation 4.10 on the preceding page, started at $r = 0$, using the C-program given in Appendix A on page 127.

simulation: jumps of -1 and $+1$, uniform distribution on the interval $[-2, 2]$, and normally distributed $\mathcal{N}(0, 1)$.

We will typically work in two dimensions because we assume the robots to move in a plane. We define $\mathbf{F}(t)$ as a normalized noise term $\|\mathbf{F}(t)\| = 1$, with mean $\langle \mathbf{F}(t) \rangle = 0$, and uncorrelated in time $\langle \mathbf{F}(t)\mathbf{F}(t') \rangle = \delta(t - t')$. The drift $\mathbf{A}$ and the diffusion coefficient B might be time- and/or space-dependent. Thus, we get the following Langevin equation which forms the first corner stone of this work:

$$\frac{d\mathbf{R}}{dt} = \mathbf{A}(\mathbf{R},t) + B(\mathbf{R},t)\mathbf{F}(t). \tag{4.11}$$

For examples of sampled two-dimensional trajectories see Figure 4.2.

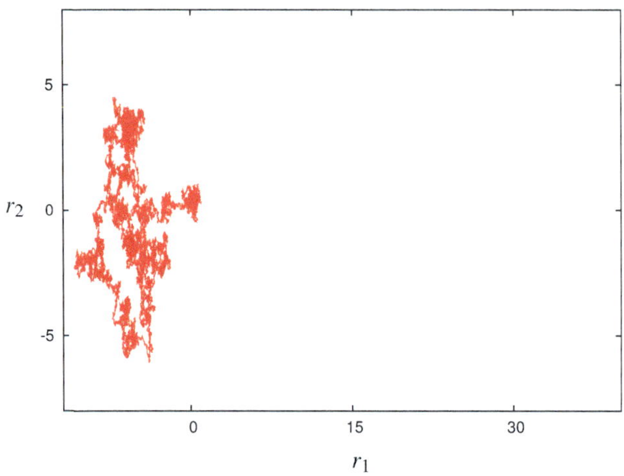

(a) One trajectory with $\mathbf{A} = (0,0)^T$ (no drift).

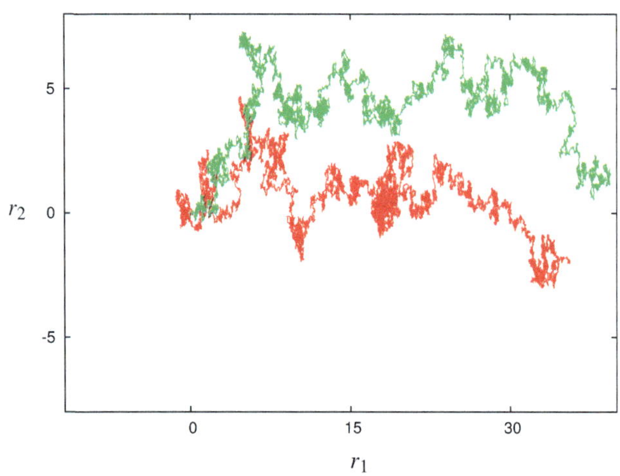

(b) Two trajectories with $\mathbf{A} = (0.1,0)^T$ (with drift in the r_1 direction).

Fig. 4.2 Sampled trajectories in 2-D without and with drift using the Langevin equation, started at $r = (0,0)^T$ and $t = 0$, stopped at $t = 500$

4.1.3 A Macroscopic Model – The Fokker–Planck Equation

Examples for derivations of the Fokker–Planck equation with the relation to the Langevin equation can be found, for example, in Kampen (1981); Haken (1977); Doob (1953). Here, we derive the Fokker–Planck equation following closely the derivation of Haken (1977) which is a great piece of technical writing. We restrict ourselves to the derivation of the one-dimensional space variant. The generalization to the two-dimensional case (actually to any dimension) uses the same methods but is hardly presentable in a comprehensible way. Nevertheless, we will discuss some selected issues of the two-dimensional case at the end of this section.

We begin by considering the trajectory in one-dimensional space of a Brownian particle in a potential. We describe this trajectory with a probability distribution using the Dirac delta function although this might seem to be unnecessary by now. The probability distribution of trajectory r_1 is

$$P_1(r,t) = \delta(r - r_1(t)), \tag{4.12}$$

for trajectory r_2

$$P_2(r,t) = \delta(r - r_2(t)), \tag{4.13}$$

and so on. Now, we average all such trajectories and introduce function

$$\rho(r,t) = \langle P(r,t) \rangle. \tag{4.14}$$

Say p_i is the probability of trajectory i then the probability distribution can be calculated by evaluating

$$\rho(r,t) = \sum_i p_i \delta(r - r_i(t)). \tag{4.15}$$

Using equations 4.12, 4.13, and 4.14 we can rewrite this and get

$$\rho(r,t) = \langle \delta(r - r(t)) \rangle. \tag{4.16}$$

The product $\rho(r,t)dr$ is the probability of encountering the particle at position r within the interval defined by dr at time t. However, we would need to consider the whole time interval to evaluate Equation 4.15 since the particle could approach position r at any time t. To avoid that, we derive a differential equation for ρ by considering the change of ρ within a time interval Δt

$$\Delta\rho(r,t) \equiv \rho(r,t+\Delta t) - \rho(r,t) \tag{4.17}$$

and using Equation 4.16 we get

$$\Delta\rho(r,t) \equiv \langle \delta(r - r(t+\Delta t)) \rangle - \langle \delta(r - r(t)) \rangle. \tag{4.18}$$

We set

$$r(t+\Delta t) = r(t) + \Delta r(t) \tag{4.19}$$

and derive the Taylor series until the second power for

$$\begin{aligned}\langle\delta(r-r(t)-\Delta r(t))\rangle =&\langle\delta(r-r(t))\rangle\\ &+\left\langle\left[-\frac{d}{dr}\delta(r-r(t))\right]\Delta r(t)\right\rangle\\ &+\frac{1}{2}\left\langle\left[\frac{d^2}{dr^2}\delta(r-r(t))\right](\Delta r(t))^2\right\rangle.\end{aligned} \tag{4.20}$$

Applying this Taylor series to Equation 4.18 results in

$$\Delta\rho(r,t)=\left\langle\left[-\frac{d}{dr}\delta(r-r(t))\right]\Delta r(t)\right\rangle+\frac{1}{2}\left\langle\left[\frac{d^2}{dr^2}\delta(r-r(t))\right](\Delta r(t))^2\right\rangle. \tag{4.21}$$

Now, we are in a position to draw the relation between this derivation of the Fokker–Planck equation and the underlying Langevin equation

$$\frac{dr}{dt}=A(r(t),t)+B(r(t),t)F(t), \tag{4.22}$$

which is needed to determine Δr. This is done by integration. We assume that the changes of r, A and B are small within Δt. Thus, we can approximate

$$\int_t^{t+\Delta t} A(r(t'),t')dt' \approx A(r(t),t)\Delta t, \tag{4.23}$$

$$B(r(t),t)\approx B(r(t+\Delta t),t+\Delta t). \tag{4.24}$$

As discussed we integrate Equation 4.22 and get

$$\int_t^{t+\Delta t}\frac{dr}{dt'}dt' = r(t+\Delta t)-r(t)\equiv\Delta r \tag{4.25}$$

$$=\int_t^{t+\Delta t} A(r(t'),t')dt' + \int_t^{t+\Delta t} B(r(t'),t')F(t')dt'. \tag{4.26}$$

While the first summand of Equation 4.26 is evaluated according to Equation 4.23 we have to treat the second summand. We define the antiderivative of F by

$$\hat{F}(t)=\int F(t')dt'\bigg|_t \tag{4.27}$$

and the derivative of B by

$$\dot{B}(r(t),t)=\frac{B(r(t'),t')}{dt'}. \tag{4.28}$$

We evaluate the second summand in Equation 4.26 using integration by parts:

$$\int_t^{t+\Delta t} B(r(t'),t')F(t')dt' = B(r(t+\Delta t),t+\Delta t)\hat{F}(t+\Delta t) - B(r(t),t)\hat{F}(t) - \int_t^{t+\Delta t} \dot{B}(r(t),t)\hat{F}(t) \quad (4.29)$$

Due to Equation 4.24 we set $\dot{B}(r(t),t) \approx 0$. Thus, the last term of Equation 4.29 vanishes. By applying Equation 4.24 a second time we get

$$\int_t^{t+\Delta t} B(r(t'),t')F(t')dt' = B(r(t),t)(\hat{F}(t+\Delta t) - \hat{F}(t)) \quad (4.30)$$

$$= B(r(t),t) \int_t^{t+\Delta t} F(t')dt' \quad (4.31)$$

$$= B(r(t),t)\Delta F \quad (4.32)$$

Putting this result and Equation 4.23 in Equation 4.26 yields

$$\Delta r = A(r(t),t)\Delta t + B(r(t),t)\Delta F(t). \quad (4.33)$$

Putting the right hand side of Equation 4.33 in the first term of the right hand side of Equation 4.21 results in

$$-\frac{d}{dr}[\langle\delta(r-r(t))(A(r(t),t)\Delta t)\rangle + \langle\delta(r-r(t))B(r(t),t)\rangle\langle\Delta F\rangle] \quad (4.34)$$

Separating $\langle\Delta F\rangle$ is approved since ΔF contains all collisions after t, but B and $r(t)$ are determined by all collisions before t, and the collisions themselves are independent from each other. Thus, the overall mean value can be expressed as a product of mean values. We assume ΔF and B to be uncorrelated. We substitute $r(t)$ by r and include $A(r,t)$ in the derivative due to subtleties in handling the δ-function in equations such as Equation 4.18 (see Haken (1977, chap. 6) for details). Now the important assumption of a quasi-equilibrium becomes prominent. We assume that the random forces cancel each other out. Furthermore, we assume that Δt was big enough such that already many collision have occurred. Thus, the average of F vanishes and, therefore, the average of ΔF vanishes as well. Hence, Equation 4.34 is reduced to

$$-\Delta t\frac{d}{dr}[\langle\delta(r-r(t))A(r,t)\rangle]. \quad (4.35)$$

Using the same arguments as before we can restate the second term of the right hand side of Equation 4.21

$$\frac{d^2}{dr^2}[\langle\delta(r-r(t))\rangle\langle(\Delta r)^2(t)\rangle]. \tag{4.36}$$

Substituting Δr by the right hand side of Equation 4.33 results in terms containing ΔF, $\Delta F \Delta t$, or $(\Delta t)^2$. The terms with ΔF vanish. The product including $(\Delta F)^2$ is the only non-zero term being linear in Δt. We get

$$\langle(\Delta F)^2\rangle \equiv \langle\Delta F(t)\Delta F(t)\rangle = \int_t^{t+\Delta t}\int_t^{t+\Delta t}\langle F(t')F(t'')\rangle dt'dt''. \tag{4.37}$$

We assume a δ-correlation for F, that is, we assume that the collisions (events disturbing the trajectory of a particle) are uncorrelated in time. Hence, we have

$$\langle\Delta F(t)\Delta F(t')\rangle = Q\delta(t-t'). \tag{4.38}$$

Now we are in a position to evaluate Equation 4.37 and we get

$$Q\Delta t. \tag{4.39}$$

Thus, we end up with a new form for Equation 4.36

$$\frac{d^2}{dr^2}[\langle\delta(r-r(t))B^2(r,t)\rangle Q\Delta t]. \tag{4.40}$$

Dividing the original (Equation 4.21) by Δt and using the two results (4.40 and 4.35) we get

$$\frac{d\rho(r,t)}{dt} = -\frac{d}{dr}(A(r(t),t)\rho(r,t)) + \frac{1}{2}Q\frac{d^2}{dr^2}(B^2(r(t),t)\rho(r,t)), \tag{4.41}$$

which is the Fokker–Planck equation describing the temporal evolution of a particle's probability distribution in one-dimensional space. As mentioned before the two-dimensional case is derived using the same methods. However, a concise representation of this derivation is hardly possible. The two-dimensional equivalent of Equation 4.16 is

$$\rho(\mathbf{r},t) = \langle\delta^2(\mathbf{r}-\mathbf{r}(t))\rangle, \tag{4.42}$$

with

$$\mathbf{r} = \begin{pmatrix} x \\ y \end{pmatrix}, \tag{4.43}$$

and $\delta^2(\mathbf{r}) = \delta(x)\delta(y)$. Then for the Taylor series (cf. Equation 4.20) we get

$$\langle\delta^2(\mathbf{r}-\mathbf{r}(t)-\Delta\mathbf{r}(t))\rangle = \langle\delta^2(\mathbf{r}-\mathbf{r}(t))\rangle + \langle(\Delta\mathbf{r}(t)\cdot\nabla)\delta^2(\mathbf{r}-\mathbf{r}(t))\rangle \tag{4.44}$$

$$+\frac{1}{2}\langle\Delta\mathbf{r}(t)\cdot[\Delta\mathbf{r}(t)\cdot\nabla(\nabla\delta^2(\mathbf{r}-\mathbf{r}(t)))]\rangle, \tag{4.45}$$

for the differential operator

$$\nabla = \begin{pmatrix} \frac{\partial}{\partial x} \\ \frac{\partial}{\partial y} \end{pmatrix}. \tag{4.46}$$

The two-dimensional Langevin equation equivalent to Equation 4.22, which will be used throughout this paper, is

$$\dot{\mathbf{r}}(t) = \mathbf{A}(\mathbf{r}(t),t) + B(\mathbf{r}(t),t)\mathbf{F}(t). \tag{4.47}$$

Concerning the correlations of $(\Delta\mathbf{F})^2$ (cf. Equation 4.38) all combinations of the components need to be considered in general

$$\langle \Delta F_x(t)\Delta F_x(t')\rangle = Q_{xx}\delta(t-t'), \tag{4.48}$$

$$\langle \Delta F_x(t)\Delta F_y(t')\rangle = \langle \Delta F_y(t)\Delta F_x(t')\rangle = Q_{xy}\delta(t-t') = Q_{yx}\delta(t-t'), \tag{4.49}$$

$$\langle \Delta F_y(t)\Delta F_y(t')\rangle = Q_{yy}\delta(t-t'). \tag{4.50}$$

However, in this work the x- and y-components are considered to be uncorrelated ($Q_{yx} = Q_{xy} = 0$), that is, a component of the strength vector caused by a collision is not influenced by other components. Hence, we set $Q = Q_{xx} = Q_{yy}$ and get

$$\frac{\partial\rho(\mathbf{r},t)}{\partial t} = -\nabla(\mathbf{A}(\mathbf{r}(t),t)\rho(\mathbf{r},t)) + \frac{1}{2}Q\nabla^2(B^2(\mathbf{r}(t),t)\rho(\mathbf{r},t)), \tag{4.51}$$

where $\rho(\mathbf{r},t)dxdy$ is the probability of encountering the particle at position $\mathbf{r}$ within the rectangle defined by dx and dy at time t. This equation is the cornerstone of our model framework which will be used throughout this paper. We restrict ourselves to the two-dimensional case but the step to a three-dimensional model would be straight forward.

Note that for $\mathbf{A} = 0$ and $B = \sqrt{2}$ Equation 4.51 is reduced to Fick's second law

$$\frac{\partial\rho(\mathbf{r},t)}{\partial t} = Q\nabla^2\rho(\mathbf{r},t), \tag{4.52}$$

which describes a standard diffusion process.

In Figure 4.3 on the next page the temporal evolution of an initial peak $\rho(0,0) \to \infty$ using the one-dimensional Fokker–Planck equation and four sample trajectories of the corresponding Langevin equation are shown (see the Figure caption for details). The C-program, that was used, is given in Appendix B on page 129.

In Figure 4.4 on the next page a similar situation for higher diffusion and the trajectory of highest probability, which is in this case just a line with slope A, are shown.

In Figure 4.5 on page 53 we compare the numerical solution of the one-dimensional Fokker–Planck equation to a histogram obtained by sampling trajectories using the corresponding Langevin equation (Monte Carlo method).

The Fokker–Planck equation belongs to the class of partial differential equations (PDE). In general, PDE are hard to solve if they are not of a standard type, which is

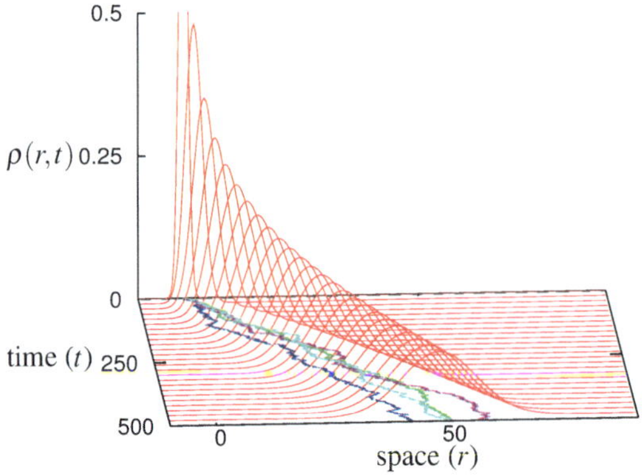

Fig. 4.3 Probability density ρ describing the probability of encountering the particle started at $t = 0$ and $r = 0$ at a given point in (one-dimensional) space computed by the Fokker–Planck equation for drift $A = 0.1$ and diffusion $B = 0.3$. On the rt-plane four trajectories are plotted computed by the corresponding Langevin equation.

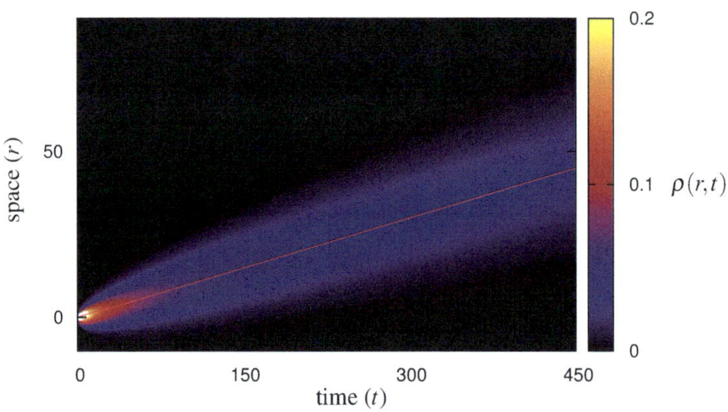

Fig. 4.4 Probability density ρ in (one-dimensional) space computed by the Fokker–Planck equation for drift $A = 0.1$, diffusion $B = 0.6$, and an initialization close to a δ-function at $r = 0$. The line ($r = At = 0.1t$) indicates the trajectory of highest probability.

not the case here. It cannot be solved explicitly except for a few special cases being of no interest in this work (Risken, 1984; Kampen, 1981). Therefore, we need to apply methods of numerical integration.

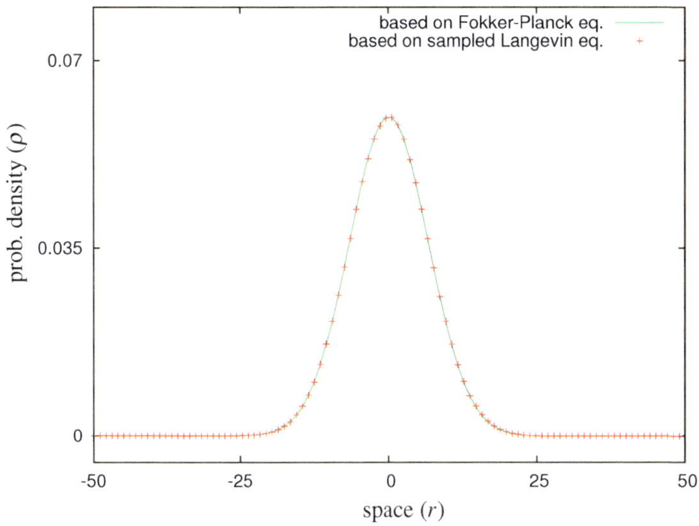

(a) $A = 0, t = 500$, based on $3 \cdot 10^6$ samples.

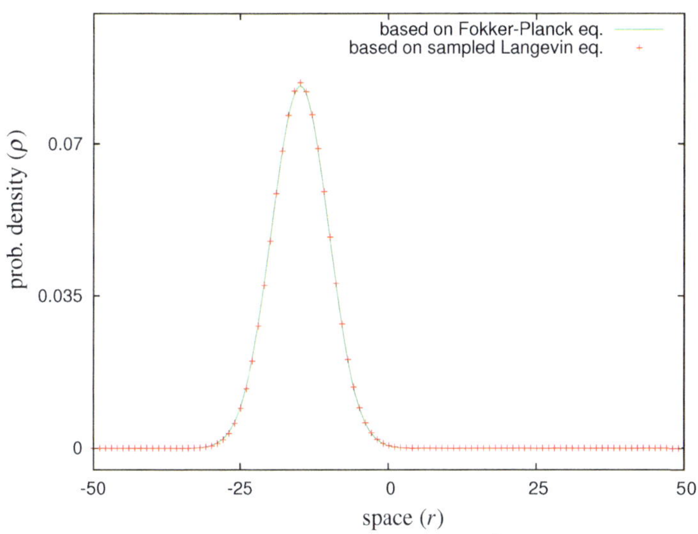

(b) $A = -0.06, t = 250$, based on $3 \cdot 10^6$ samples.

Fig. 4.5 Histograms of sampled trajectories using the Langevin equation in comparison with the solution to the corresponding Fokker–Planck equation for initial peaks at $r = 0$ and $B = 0.3$, 95% confidence intervals are within the symbols' size.

Discretizing

The first step, before numerical algorithms are applicable is to discretize the equation. The three most important methods are: finite differences, finite volumes, and finite elements. Here, we will use the finite difference method. Time is transformed to discrete time steps and space is discretized by a finite grid.

The following three differences are the building blocks for discretizing the Fokker–Planck equation: the first-order forward difference for the time derivative at time n and point ξ in one-dimensional space

$$\frac{\partial u_\xi^n}{\partial t} \doteq \frac{u_\xi^{n+1} - u_\xi^n}{\Delta t}; \tag{4.53}$$

the first-order central difference for the space derivative

$$\frac{\partial u_\xi^n}{\partial x} \doteq \frac{u_{\xi+1}^n - u_{\xi-1}^n}{2\Delta x}; \tag{4.54}$$

and the second-order central difference for the space derivative

$$\frac{\partial (u_\xi^n)^2}{\partial x^2} \doteq \frac{u_{\xi+1}^n - 2u_\xi^n + u_{\xi-1}^n}{(\Delta x)^2}. \tag{4.55}$$

Once the equation is discretized we transform it to a linear system of equations and then need an efficient method to solve it. This is discussed in the next section.

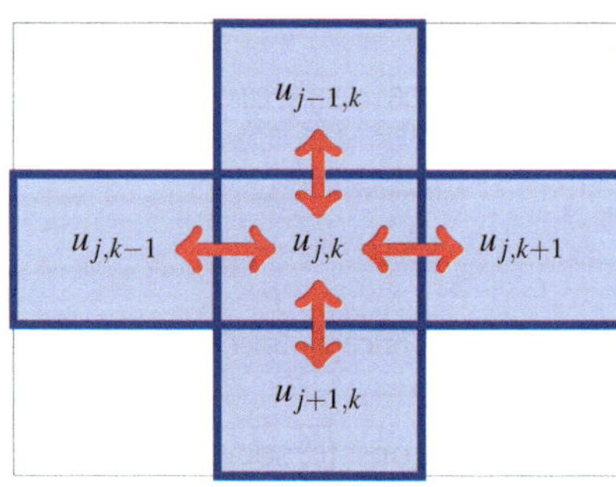

Fig. 4.6 Schematic diagram of Equation 4.58 focusing on one patch in the center and its Von Neumann neighborhood. Arrows indicate the in- and outflow.

Solving

In the following we restrict ourselves to the treatment of initial value problems only, that is, problems with given values for one point in time and we are interested in the evolution forward in time. The simplest way of solving the discretized equation is the forward integration in time. This method just implements the finite differences directly by summing over the time steps.

However, this method has a poor stability, that is, errors due to the limited accuracy of the floating point arithmetic might successively be magnified. Instability is one of the main issues in numerics (Higham, 2002). To overcome the problem of instability in case of the forward integration only small step sizes in space and time are allowed. However, this makes many iterations necessary to calculate the evolution in time.

Fortunately, numerics provides us with more sophisticated methods to speed this process up and staying stable. We use the Crank–Nicholson method (Crank and Nicolson, 1947) combined with the alternate-direction implicit method (ADI). The Crank–Nicholson method is based on the central difference in space, and the trapezoidal rule in time, that is, it is implicit. A numerical method is implicit if the value being determined occurs also on the right hand side of the equation. The Crank–Nicholson method for two dimensions in space and time step size of Δt

$$u_{j,k}^{n+1} = u_{j,k}^{n} + \frac{\Delta t}{2}\left(G_{j,k}^{n+1} + G_{j,k}^{n}\right), \tag{4.56}$$

with G_j^n is the discretization of the PDE's right hand side or of a single operator respectively, for example in case of the two-dimensional diffusion equation

$$\frac{\partial u}{\partial t} = D\left(\frac{\partial^2 u}{\partial x^2} + \frac{\partial^2 u}{\partial y^2}\right), \tag{4.57}$$

we get in a uniform grid with grid point distances of Δx

$$G_{j,k}^{n} = \frac{D}{(\Delta x)^2}\left(u_{j+1,k}^{n} + u_{j-1,k}^{n} + u_{j,k+1}^{n} + u_{j,k-1}^{n} - 4u_{j,k}^{n}\right). \tag{4.58}$$

See Figure 4.6 on the preceding page for a visualization of Equation 4.58.

Equation 4.56 with G defined by Equation 4.58 written in matrix notation results in a matrix with five non-zero diagonals. Less non-zero diagonals would be desirable, for example, a tridiagonal matrix is disproportionately easier to solve.

This can be achieved by ADI which is an operator splitting method reducing the system of equations to an easier to solve problem. For example, the two-dimensional diffusion equation splits into two equations

$$u_{j,k}^{n+\frac{1}{2}} = u_{j,k}^{n} + \frac{D\Delta t}{2(\Delta x)^2}\left(u_{j+1,k}^{n+\frac{1}{2}} - 2u_{j,k}^{n+\frac{1}{2}} + u_{j-1,k}^{n+\frac{1}{2}} + u_{j,k+1}^{n} - 2u_{j,k}^{n} + u_{j,k-1}^{n}\right), \tag{4.59}$$

$$u_{j,k}^{n+1} = u_{j,k}^{n+\frac{1}{2}} + \frac{D\Delta t}{2(\Delta x)^2}\left(u_{j+1,k}^{n+\frac{1}{2}} - 2u_{j,k}^{n+\frac{1}{2}} + u_{j-1,k}^{n+\frac{1}{2}} + u_{j,k+1}^{n+1} - 2u_{j,k}^{n+1} + u_{j,k-1}^{n+1}\right). \tag{4.60}$$

Finally, we have to define the boundary conditions and the initial values of the grid for $t = 0$. Here, we have Neumann boundary conditions, that is, they are defined by the derivative on the boundary of the domain. For the case of a totally isolated boundary, that is, no agent will ever leave the domain, we set

$$\frac{\partial u}{\partial \nu}(\mathbf{x}) = 0, \quad \forall \mathbf{x} \in \partial\Omega, \tag{4.61}$$

for the boundary $\partial\Omega \subset \mathbb{R}^2$ and ν denotes the normal to the boundary. Then we get, for example, for Equation 4.59 and a grid point (j,K) in an outest column K

$$u_{j,K}^{n+\frac{1}{2}} = u_{j,K}^{n} + \frac{D\Delta t}{2(\Delta x)^2}(u_{j+1,K}^{n+\frac{1}{2}} - 2u_{j,K}^{n+\frac{1}{2}} + u_{j-1,K}^{n+\frac{1}{2}} - u_{j,K}^{n} + u_{j,K-1}^{n}). \tag{4.62}$$

The system of equations defined by equations 4.59-4.62 can simply be expressed using tridiagonal matrices (the values of the grid points are represented by a single, huge vector $\mathbf{u}$)

$$\mathbf{M}\mathbf{u}^{n+1} = \mathbf{N}\mathbf{u}^{n}, \tag{4.63}$$

with

$$\mathbf{M} = \begin{pmatrix} -1 & 1 & & & & & & & \cdots & 0 \\ 1 & -2 & 1 & & & & & & & \vdots \\ & & \ddots & & & & & & & \\ & & 1 & -2 & 1 & & & & & \\ & & & 1 & -1 & & & & & \\ & & & & & -1 & 1 & & & \\ & & & & & 1 & -2 & 1 & & \\ & & & & & & & \ddots & & \\ \vdots & & & & & & & 1 & -2 & 1 \\ 0 & & \cdots & & & & & & 1 & -1 \end{pmatrix}, \tag{4.64}$$

and $\mathbf{N}$ accordingly. For a $j \times k$ grid we have jk unknowns and jk equations. By defining the initial values of the grid $u_{j,k}^{0}$ we complete the mathematical description of the initial value problem and can start to solve it.

The linear system of Equation 4.63 is represented by sparse matrices in the computer. It is solved by fast iterative solvers such as Jacobi, Gauss–Seidel, successive over-relaxation, or conjugate gradients. We do not specify any of these methods since, once we have reduced the original equation to a linear system of equations, several out-of-the-box tools (e.g., Matlab™, Maple™) can be used to solve it. However, these standard tools do not provide an automatic discretization of non-standard PDE depending not only on constant coefficients but might even be coupled to other PDE.

4.2 Discussion of the Methodology

In the following we discuss how a model can be applied in the design of swarm algorithms and why the application of a model is suggestive.

4.2.1 *Purpose of a Model in the Design of Swarm Algorithms*

The direct approach in programming a robotic swarm is insufficient as discussed in Section 3.2.3 on page 36. The proposed approach is to overcome these challenges by supporting the design process with models. The basic model-based approach in contrast to the direct approach is shown in Figure 4.7. The algorithm designer needs to find a conception of the expected swarm behavior while creating this behavior from scratch. However, one is deluded by oneself's intuition and the actually resulting behavior is unexpected. The solution, proposed here, is to support the design phase by a model delivering accurate predictions that correspond well to reality. This way the designer gets an early feedback to the current algorithm and is able to adapt it in an iterative process.

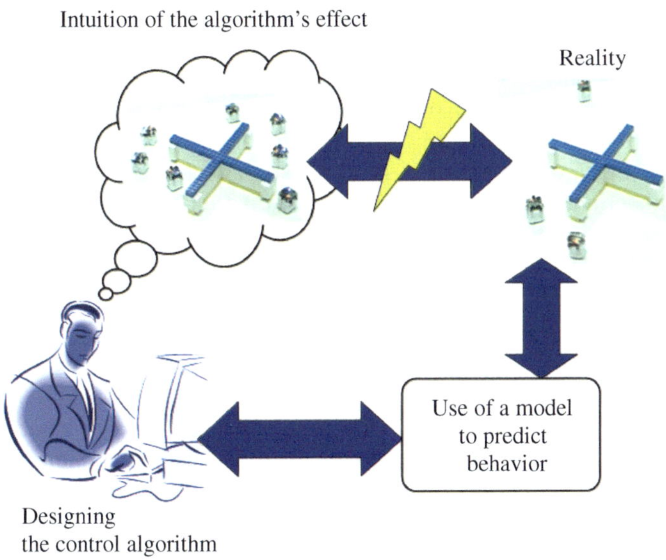

Fig. 4.7 Schema comparing direct programming with the model-based approach of designing swarm algorithms

In Figure 4.8 on the next page a schema is given that shows the basic model-based approach. We assume that a task, which has to be solved by a robotic swarm, is preassigned as well as the hardware. The following implementation step is different from the classical software engineering approach. Based on a global description we have to find and define an algorithm for a single component based on local information only. This is sometimes called global-to-local programming , see Section 3.2.1 on page 35 (Yamins, 2007). At a next step the basic design of a first control algorithm is done. Typically, such a swarm algorithm will have several parameters that significantly define the resulting behavior. By now we have two constraints: One is the defined task and the other one is the basic algorithm as it restricts the achievable global behavior (c.f. Figure 4.8 on the next page).

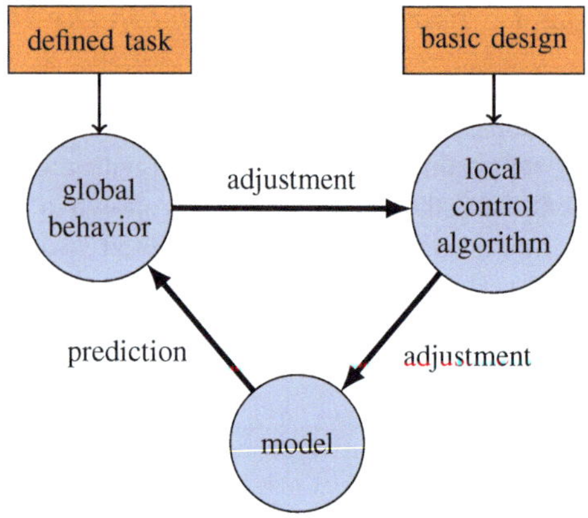

Fig. 4.8 Schema of applying models in designing swarm algorithms

In the following, we assume that the set of possible behaviors defined by the task and those defined by the basic algorithm overlap, that is, the task can in principle be solved or at least approximately be solved by the basic algorithm. The actual control algorithm and the parameter configuration is, however, still unknown. Based on the basic algorithm a model is designed, which incorporates these parameters ideally directly. Otherwise, a mapping has to be defined that maps the algorithm parameters to the model parameters and vice versa. Another property of the algorithm might be a form of modularity that might lead to a modularized model as well and simplify the model design.

An accurate model will predict the global behavior, which is to be expected based on the given algorithm. Based on these predictions the parameters of the algorithm or the algorithm itself are adjusted. These adjustments need to be done for the model, too. Now, a second iteration starts by using the model again to gain predictions of the current algorithm. This is repeated until the global behavior as predicted by the model is satisfying. This evaluation of the behavior can be done with a defined formal metric (e.g., robot densities in important areas) or informally (e.g., subjective estimation of emerging patterns). In an ideal case the performance evaluation could be based on parameters only. If the model can automatically be adapted to new parameters and there is a computable metric the whole process can be automated to a search for an optimum in the parameter space of the algorithm. However, typically the adjustments to the model are rather complex and need a human model designer. Furthermore, some simulation runs or even experiments with a few robots might be necessary to measure macro-level parameters and to adapt the model.

4.2.2 Benefits of a Symbolic Model

An agent-based simulation might seem to be the easier approach to analyze swarms and indeed the implementation of a simulation is straightforward. Such a simulation can be advanced as close to reality as desired without any need for abstractions. However, the possibilities of analysis are narrowed down to mere measurements.

A model might offer more methods of analysis, especially a symbolic model that allows analytic manipulations. A good motivation for the modeling approach in agent systems is given by Schweitzer (2003):

> To gain insight into the interplay between microscopic interactions and macroscopic features, it is important to find a level of description that, on the one hand, considers specific features of the system and is suitable for reflecting the origination of new qualities, but, on the other hand, is not flooded with microscopic details. In this respect, agent models have become a very promising tool mainly for simulating complex systems. A commonly accepted theory of agent systems that also allows analytical investigations is, however, still pending because of the diversity of the various models invented for particular applications.

We identify three potential benefits of a symbolic model: they are used to gain insight, they are easy to display, and they abstract from disturbing details:

1. Symbolic models lead to insights due to their explanatory ability unlike numerical models or simulations. In a simulation the swarm behavior can only be observed. Using a symbolic model the methods of analysis are manifold and by designing the model, assumptions have to be adopted that allow further theoretical implications.
2. The representation of symbolic models is concise and complete. They are easy to compare if displayed as closed-form expression. For example, if models A and B have one term in common but model B has an extra term as well and shows a slightly different behavior for the same parameters then this difference originates in the additional term of model B. In contrast, simulations are typically only completely described by their software implementation, which is difficult to display and to compare.
3. As mentioned above, there is a demand of representing important macroscopic features while ignoring irrelevant details on the micro-level. The necessary reduction of microscopic details can most likely only be achieved by a probabilistic approach and through abstraction. The exact prediction of a robot's position is not only ruled out by the chaos-theoretic argument, stating that the initial setting can never be measured with sufficient accuracy, but also by involved processes being widely considered to be probabilistic in principle, for example, wireless communication or noise in sensory information. Furthermore, significant sources of complexity are the agent-agent interactions. The complexity caused by interactions is overcome by omitting an explicit agent representation.

In summary, it can be said that a symbolic model offers much more explanatory power compared to simulations. This benefit is gained by a necessary extra effort that needs to be invested in the model design phase.

4.2.3 Consideration of Alternative Methods

The set of methods that are available to model swarms is certainly huge. Thus, an exhaustive consideration of all possibilities is difficult. We try to get close to a comprehensive but very concise discussion by treating classes of methods and by giving a few prominent examples.

The first step in choosing our methodology is to decide between an algorithmic (simulation) and a symbolic (analytic) approach. Most of the state-of-the-art approaches in swarm robotics are algorithmic because simulations are dominating. Usually algorithmic means also microscopic. However, this is not inevitable as also macroscopic approaches can be implemented in an algorithmic fashion. Extensions to standard numerical algorithms, for example, by extensive if-then-constructs are feasible that go beyond what is analytically representable today. The problem of all algorithmic approaches is that a concise presentation is hardly possible. However, that is one of our objectives in this work.

The second step in choosing the methodology is to decide between a discrete or continuous space representation. We immediately discard all models, that represent space only explicitly in a discrete way (e.g., by defining macroscopic and application-specific patches, cf. Martinoli (1999)) because one of our objectives is the capability of representing inhomogeneous space in a straightforward manner. This leaves only models representing space in a standard, usually a regular way (e.g., square or hexagonal grids). One example of a good candidate is the class of lattice gas automata. However, in this work we want to have a concise representation of the model, we want to limit ourselves as little as possible, and we want to emphasize the micro-macro link. Therefore, we choose continuous space as it potentially meets these requirements.

The third step is to decide between discrete or continuous time. One example of a good candidate with discrete time would be the Chapman–Kolmogorov equation. However, we choose – as above – the continuous approach in order not to restrict ourselves. Thus, an analytical approach is still feasible even though maybe more complex than time-discrete approaches. Incidentally, the effect of this discrimination is relevant in the analytical approach but of limited importance in the actual implementation. The implementation of the model will be numerical in this work due to the limited possibilities of analytical solutions. Therefore, we use ordinary differential equations and partial differential equations in this work.

4.3 A Physically Motivated Approach

We want to apply the Langevin and the Fokker–Planck equation as presented and derived in Sections 4.1.2 on page 42 and 4.1.3 on page 47 to robot swarms. However, our robots are autonomous in contrast to natural particles that just react to external forces and do not decide on their own what to do next. In addition, these equations have been developed to investigate a single particle showing Brownian motion and not a large group of such particles. In the model of Brownian motion with drift there are two classes of particles: The single, focal big particle showing some deterministic and some random motion and the many particles, in which the big particle is floating, that are smaller by about one magnitude. Only the small particles' effect to the bigger particle is modeled and not their actual trajectories. Here in the case of swarm robots, we have just one class of particles/agents: We focus on a homogeneous robot swarm. The nondeterministic component of their motion is caused by many factors including the rest of the swarm.

A second major difference is the possibility of cooperation by means of direct or indirect communication. The robots can decide to explicitly change the environment in a way that bears information for other robots. Unfortunately, there is no theory in physics on explicitly communicating particles. Thus, we have to find a viable way of extending the model appropriately.

This physically motivated approach combined with empirically motivated extensions serves as a construct of ideas. The metaphorical allusion to well-known processes in physics is an inspiration to the model designer, just as biology is an inspiration to the algorithm designer.

A partial solution of the micro-macro problem (see Section 2.1.8 on page 9) is implied by the fact that at least the robots' motion (typically not the complete behavior) can be described by a "standard" Langevin and the corresponding Fokker–Planck equation. From the microscopic description (Langevin equation) we are able to derive the macroscopic description (Fokker–Planck equation).

In the following section the motion model is presented, which is based fully on the Langevin and the Fokker–Planck equations. In the successive section an extension of the model to incorporate communication is discussed.

4.4 Modeling Robot Motion

In the following we try to find the causes for deterministic motion of swarm robots and the causes for nondeterministic motion. This is important for our model in order to develop well defined functions $\mathbf{A}$ and B. In swarm robotics, the motion of a robot is typically categorized in two types: First, the robot performs a type of random walk. Second, it moves in a direction determined by processing information provided by the environment. In general, there are three causes for random motion (cf. Figure 4.9 on the next page): One, there is no information available to the robot; thus, the robot has to move randomly. Two, information is available but the autonomous robot decides to move randomly instead of exploiting this information. Three, information is available and the robot would like to move goal-oriented but due to physical constraints, for example, another robot or an obstacle is blocking its way, a random direction needs to be chosen to keep moving and to avoid deadlocks.

Now, we consider the possibilities and challenges of modeling these classes of random and goal-oriented motion. In the case of lacking information, it is necessary to know the potential source. If the lack of information is due to missing features in the environment it can be modeled with the help of the environment's model as discussed below in Section 4.5.1 on page 65. If the lack of information is due to a missing message from another robot a model of communication is applied (see Section 4.5.2 on page 66). If the lack of information is due to a missing neighbor this can be modeled by the the probability of the event "there is no neighbor". The Fokker–Planck equation delivers this probability directly. At this point, we are

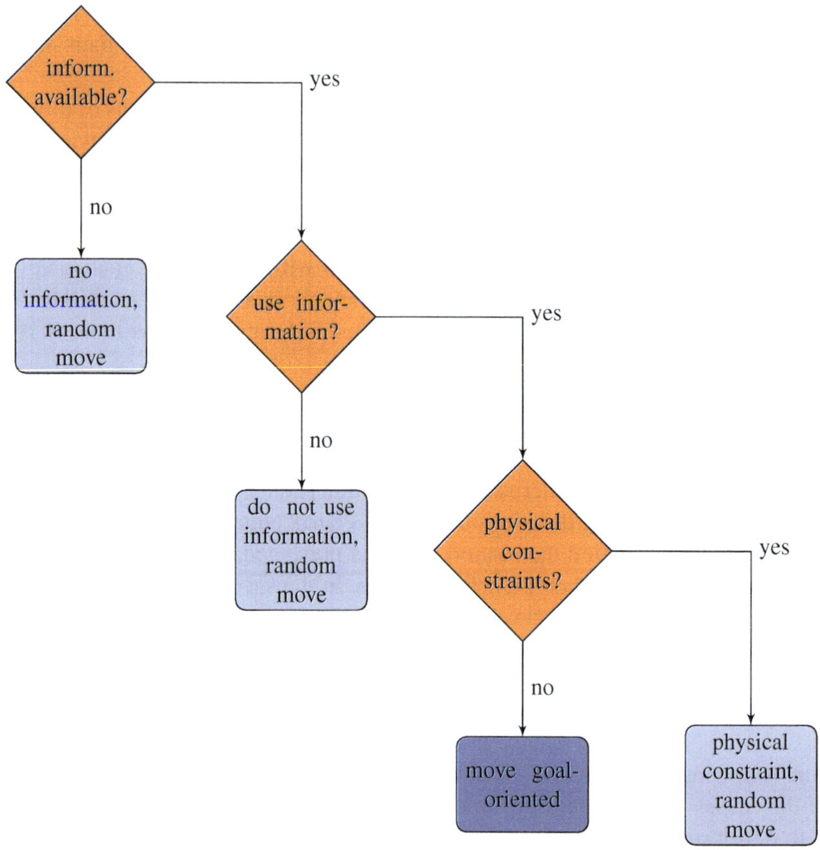

Fig. 4.9 Schema, classes of random moves (Hamann and Wörn, 2008b)

approaching an unbalance between the Langevin and the Fokker–Planck equation. Using the Langevin equation we work on explicit trajectories. Hence, a probability density is not defined or only on the basis of delta functions. Thus, we either evaluate the actual present situation, that is, we deterministically determine whether there is a neighbor, or we use a probability distribution that was sampled in a bootstrap process. This will be discussed in Section 4.8.1 on page 72. In the following, we mainly focus on the Fokker–Planck equation in such unbalanced situations.

In the case of available information being not used due to an autonomous decision of the robot, the underlying exploration rate can be extracted from the control algorithm and can be incorporated in the deterministic and indeterministic motion term of the model.

In the case of physical constraints (collision-avoidance behavior), the way of modeling depends, as in the case of lacking information above, on the cause of this constraint, which is either a feature of the environment (see Section 4.5.1 on page 65) or another robot.

Finally, all classes of random motion will be modeled by a diffusion term, although the methods of deriving the appropriate coefficients differ in dependence on the actual effect. The robots' collision avoidance behavior, for example, shows features that are well modeled by diffusion processes: From time to time, the robots' movements are interrupted by encountering each other. This event can metaphorically be interpreted as colliding particles because both robots will change their direction after a collision. However, an actual collision is avoided because the robots are programmed to turn away from each other before getting too close. This turning angle can be considered to be random due to the complex underlying processes in the sensors and due to the robots' imprecision. Thus, robots in dense areas collide many times until they have chosen a direction by chance that leads them out of the pile. Hence, in correspondence to a diffusion process, robots tend to move away from areas, which are crowded with robots, into un-populated regions. Concerning the other classes of random motion, we assume that the chosen directions are uncorrelated in time and uniformly distributed. Using these assumptions diffusion is a well description of the random motion component of swarm robots taking enough collisions. However, with decreasing mean robot density and decreasing number of collisions this approach becomes, in principle, more and more inaccurate. This is true except for the special case that the deterministic motion leads to a uniform distribution of the robots as well, for example see Section 5.1 on page 77. Then the collisions are indeed not the cause of this homogenization process but a diffusion process might still be an appropriate model option (for a discussion of this issue see Section 4.8 on page 71).

In the following we do not yet discuss the source of information available to a robot helping to determine its goal-oriented motion — this problem will be explored in the next sections.

We assume that our swarm robots move at nominal speed at all times and, therefore, disregard acceleration. This is a reasonable simplification, as the inertia is minimal for the typical swarm robot compared to the engine power which results in acceleration times below one second (Jasmine, 2008; Caprari et al., 1998). The motion of such a single robot is described by a Langevin equation of its position $\mathbf{R}$ (reprinting Equation 4.11 on page 46)

$$\dot{\mathbf{R}}(t) = \mathbf{A}(\mathbf{R}(t),t) + B(\mathbf{R}(t),t)\mathbf{F}(t) \qquad (4.65)$$

with a random dislocation $\mathbf{F}$ due to fluctuating influences, which is defined as in Section 4.1.2 on page 42. Function $\mathbf{A}$ is a direction and describes the deterministic motion based on information provided by the environment (also information indirectly provided by other robots via the environment). Function B describes the random component of the motion. $\mathbf{A}$ and B are characterized by the underlying control algorithm. B typically incorporates influences by the other robots that disturb the robot, for example, by the need of collision avoidance. $\mathbf{A}$ might incorporate an external influence such as a light gradient. Typically, it also incorporates a positive influence by other robots, for example, communicated directional information. Depending on the actual modeling of $\mathbf{A}$ and B, Equation 4.65 on the preceding page might allow the isolated simulation of a single robot. However, this is only possible if there is no cooperation at all or the cooperation can be abstracted such that the influence by other robots can be estimated without simulating them.

Using a computer and a random number generator the motion of such a robot can be simulated (see Section 4.1.2 on page 42). The result is one trajectory out of the infinite set of all potential trajectories. However, to deduce the performance of a control algorithm, we are interested in the average behavior of the swarm. This behavior could be identified by sampling many trajectories, for example, by Monte Carlo simulations of the Langevin equation for the needed number of coexisting agents. However, this approach is microscopic and suffers from the same problems as standard robotic simulations, for example computational cost.

We want to discard this computational overhead by introducing a macroscopic model. From the Langevin equation the Fokker–Planck equation can be derived analytically (see Section 4.1.3 on page 47). The solution of a Fokker–Planck equation is relatively fast on today's desktop computers (typically within one minute) although this has to be done numerically (see Section 4.1.3 on page 54). We reprint Equation 4.51 on page 51:

$$\frac{\partial \rho(\mathbf{r},t)}{\partial t} = -\nabla(\mathbf{A}(\mathbf{r},t)\rho(\mathbf{r},t)) + \frac{1}{2}Q\nabla^2(B^2(\mathbf{r},t)\rho(\mathbf{r},t)), \tag{4.66}$$

where Q is the displacement by a collision. $\rho(\mathbf{r},t)dr_x dr_y$ is the probability of encountering a robot at position $\mathbf{r}$ within the rectangle defined by dr_x and dr_y at time t. Note that we use $\mathbf{R}$ for positions of individual robots and $\mathbf{r}$ for points in space not directly connected to an explicit robot. Equations 4.65 on the page before and 4.66 are the two cornerstones of our model framework which will be used throughout this work. We restrict ourselves to the two-dimensional case but the step to a three-dimensional model is straightforward.

Following Prigogine (1997), the probability density ρ can be interpreted as a superposition of trajectories:

> The probability [...] corresponds simply to a superposition of trajectories, and leads to no new properties. The two levels of description, the individual level (corresponding to single trajectories) and the statistical level (corresponding to ensembles), would be equivalent.

He even sees the probabilistic approach as the more comprehensive one (Prigogine, 1997):

> The probability distribution takes into account the complex microstructure of the phase space. The description of deterministic chaos in terms of trajectories corresponds to an over-idealization and is unable to express the approach to equilibrium.

The Fokker–Planck equation implements the necessary abstraction of microscopic details as described above and treats rapidly-changing parameters as noise. The equation is still exact if this noise is generated by a Gaussian process, that is, if it is fully determined by the first two moments. It gives the temporal evolution of the probability density describing the positions of the agents. The derivation of the Fokker–Planck equation for a given Langevin equation offers the possibility of analytically deriving some macroscopic aspects of a robot swarm out of a microscopic description.

The main features of the agents' movements are captured by Equations 4.66 on the preceding page and 4.65 on page 63 as discussed above. In order to model a swarm of interacting robots, essential extensions have to be made. Appropriate functions $\mathbf{A}$ and B need to be defined. These have also to be coupled with equations that model the processes taking place in the environment.

4.5 Modeling Robot-Robot Interactions

In the preceding section we did not discuss the origin of the information which enables a robot to move goal-oriented. This origin could be a physically defined gradient, for example, heat or natural gas, which is relatively easy to model. However, in applications of swarm robotics there is typically some form of cooperation making communication necessary. Thus, the modeling framework based on the Fokker–Planck equation needs to be extended to provide a way of modeling communication between robots. Because in physics there is no model of particles which communicate by means of a defined protocol, we have to drop the particle analogy in this section. However, the standard equations (Equations 4.65 on page 63 and 4.66 on the facing page) remain as the matrix of our model because they correctly describe the fundamental effects of motion independent of external information determining the robot's goal-oriented movements. First, we discuss how an external influence to the robot's motion (that might be influenced by the swarm as well) can be modeled in general. Second, we explain why modeling direct communication is a challenging task.

4.5.1 External Influence – Modeling the Environment

The general approach to incorporate external influence in the robot's movements is defining $\mathbf{A}$ as a function of a potential field P describing features of the environment (e.g., light intensities or pheromones). In the following, the use of both mental models of function $\mathbf{A}$ as a component of the Langevin equation or of the Fokker–Planck equation will lead to the same results. For example, we can define $\mathbf{A}$ using the gradient of P

$$\mathbf{A}(\mathbf{r},t) = -\nabla P(\mathbf{r},t). \tag{4.67}$$

By this definition we model a gradient ascent, that is, the robots aggregate at the maxima of P. Potential P can be temporally and spatially variant and we can also couple the temporal evolution of P with the density ρ. Thus, there is an extensive variety of possible applications: physically-defined processes (e.g., diffusion, heat flow, fluid dynamics, propagation of electromagnetic waves) but also heuristic approaches of possibly macroscopic effects (e.g., message propagation). Additionally, a discount coefficient should be included, especially in the case of a gradient ascent, making sure **A** drops with increasing density. Otherwise, the robots could aggregate in arbitrarily high densities.

The possibilities of defining a temporal evolution of P are manifold. In the case of physically well-defined processes, this definition can be easy. For example, for a simple diffusion process we get

$$\frac{\partial}{\partial t}P(\mathbf{r},t) = D\nabla^2 P(\mathbf{r},t), \tag{4.68}$$

for some diffusion coefficient D. Indirect communication (e.g., stigmergy) can be modeled by extending Equation 4.68. In a simple case where robots permanently emit material that diffuses and evaporates thereafter (e.g., pheromones), an additive term proportional to the robot density and a subtractive term proportional to the potential are added to Equation 4.68:

$$\frac{\partial}{\partial t}P(\mathbf{r},t) = D\nabla^2 P(\mathbf{r},t) + \gamma\rho(\mathbf{r},t) - \theta P(\mathbf{r},t), \tag{4.69}$$

for constants γ and θ. The disposal of material could be space- and time-dependent leading to a function $\gamma(\mathbf{r},t)$. As mentioned above, the possibilities are manifold and modeling them is quite intuitive.

Finding an appropriate model for more sophisticated processes such as the communication of messages (e.g., point-to-point, gossiping) is more problematic. This problem is discussed in the next section.

4.5.2 Challenge of Modeling Communication

The problems of modeling and simulating communicating robots are closely related to those experienced in the field of ad hoc and sensor networks. Most of the work in this field is based on simulations; for example, using ns-2 (McCanne et al., 1997). Analytic results are rare and the ones which exist are often only valid for special cases, for example, the one-dimensional case. For example, this applies to the problem of determining the probability of a successful (multi-hop) message transfer between two points for a given node density — called *path probability* $P(\text{path})$ by Bettstetter (2004). He states:

> Reviewing the literature on *ad hoc* networks, it seems that an analytical approach for calculating $P(\text{path})$ does only exist for one dimension and an infinitely large system space. Since we are mainly interested in two-dimensional bounded areas, we approach the problem by means of simulation.

Here, simulation means sampling random topologies in the tens of thousands for uniform distributions, not considering non-uniform distributions. Combined with a numerical solver of our model, this method is infeasible since we would either have to draw thousands of samples per time step or to rely on statistically insignificant data. Therefore, we have to approximate the effects of communication heuristically, that is, macroscopically, and in closed-form. Obviously, this approximation will be one of the main challenges in modeling scenarios which include communicating robots.

4.6 Towards General Methodological Principles

It seems unlikely that a complete step-by-step manual for designing swarm models independent of the actual application is possible. However, the proposed framework seems to have the potential to be applicable to quite a wide variety of scenarios. In this section we present a few principles for the development of a model based on the Langevin and Fokker–Planck equation. These principles are stated by generically answering three questions which should be asked by the model designer. Still, the model development will remain a partially creative process which cannot be generalized for all kinds of application.

The first question is: Are there several states in the controller and should all of them be explicitly modeled? In principle, one has to introduce a Langevin and a Fokker–Planck equation for each state m. Thus, one gets a set of Langevin equations

$$\dot{\mathbf{R}}(t) = -\mathbf{A}_m(\mathbf{R}(t),t) + B_m(\mathbf{R}(t),t)F(t) \tag{4.70}$$

plus a formalism administrating the transitions between states. The corresponding set of Fokker–Planck equations is:

$$\frac{\partial \rho_m}{\partial t} = -\nabla(\mathbf{A}_m\rho_m) + \frac{1}{2}\nabla^2(B_m^2\rho_m) + \sum_{n\neq m} t_{n,m}\rho_n - t_{m,n}\rho_m, \tag{4.71}$$

for transition rates $t_{n,m}$ which define the probability of switching from state n to m. Note the similarity to the master equation and, hence, the similarity to the rate equation approaches (Lerman et al., 2005) introduced by the sum handling the state switches. For an example of such a multi-state approach see Section 5.4 on page 105 and Hamann and Wörn (2007a). However, it is not always necessary to model each state explicitly as shown in Sections 5.1 on page 77 (collision-based aggregation) and 5.3 on page 100 (emergent taxis). The unification of states is generally possible, if the state-dependent differences of the $\mathbf{A}$- and B-functions for each state can be expressed, for example, by a spatiotemporal function (e.g., the s-function in the emergent taxis scenario, Section 5.3 on page 100).

The second question is: When do the robots move randomly and when do they move in a goal-oriented manner? If these ratios are expressible in scalar functions, the ratios are incorporated in B and as a part of $\mathbf{A}$. They might depend on external influences as well, which leads to the next question.

The third question is: Based on what information do the robots move in a goal-oriented manner? The information should be a direction which is incorporated in $\mathbf{A}$. Typically, the information depends on a natural spatiotemporal process. Such processes have models well-known from physics. If high-level communication between the robots is used one has to introduce empirical parameters and/or functions as discussed above. However, it is also possible that there is only exploration, making **A** obsolete.

4.7 Relevance to the Concept of Computation

Besides the effort of developing a model of swarm behavior that allows an efficient algorithm design process, a different idea might prove to be relevant here. In this section we discuss the possibility of interpreting swarm behavior as a form of computation following the concept of "world-embedded computation" as proposed by Payton et al. (2001). In turn, this would imply that the model proposed in this work is also a model of computation. Absorbing properties of such an interpretation of computation are:

- It is a quintessentially stochastic concept of computation and therefore very different from the classical model.
- It is an embodied concept of computation and thus very visual and describable in equations of motion.
- It is based on local interactions and local information processing only allowing a maximum of robustness.

The following disquisition was previously published in Hamann and Wörn (2007c).

4.7.1 From Swarm Behavior to World-Embedded Computation

With the increase of interdisciplinary research new concepts were developed in the last decades. For example, swarm intelligence (see Section 2.2 on page 9) applied to computational problems leads to powerful meta-heuristics (see Section 2.2.2 on page 10) and applied to robotics it results in large-scale distributed robotic systems (see Section 2.3.2 on page 12). In the following we try to pursue these approaches and to combine swarm intelligence with robotics and heuristics.

The scientific approach to computation was significantly governed by the computational devices used in the past. Thus, it began with sequential devices, later parallel machines with shared memory were studied, and even later the focus was on distributed but fully connected systems. All these approaches have determinism and explicit communication in common.

A new philosophy is introduced by applying concepts of swarm intelligence, such as simple local rules, indirect communication (stigmergy), and cooperation. We can consider swarm intelligence as the final step of the process of getting away from centralized and deterministic systems towards fully distributed and probabilistic systems. Swarm intelligence was applied to computational problems by using software agents. We provide our computational devices with actuators making them mobile. Hence, they become real, embodied agents in the form of robots. The idea of using a group of autonomous agents as processing elements, that are embedded in the environment, that sense and compute based only on local information was published by Payton et al. (2001) and propagated as "world-embedded computation" (see Section 2.6.2 on page 24). This is related to an old question, whether an ant colony's struggle of survival might be viewed as computation (Hofstadter, 1979). Thus, we note that problem solving by adaptive and cooperative behaviors might be considered computation.

Combining the local and randomized approach of swarm intelligence with an emphasis on positional information, results in an interesting computing paradigm or also in a method of generating emergent behavior. The position of a robot in the physical world becomes the building block of collective information processing. In a recent work Litus et al. (2007) give a good summary of this idea:

> The key insight that underlies our methods is that the physical locations of the robots themselves could be considered as an approximate solution to the entire problem. An individual robot can move itself, thus refining the current solution approximation. No representation of the problem, or the current solution, needs to be held by any robot: they manifest the solution by their physical configuration.

Although these approaches suffer in principle from the same problems of intractability as classical approaches concerning hard problems, they might lead, nevertheless, to more efficient implementations. Such systems might be cheaper than classical devices and easier to maintain due to less complexity.

4.7.2 The Concept of Embodied Computation

The meaning of computation according to WordNet is "determining something by mathematical or logical methods" or "problem solving that involves numbers or quantities" (WordNet 3.0, Princeton University, 2007). However, our concept denies partly the involvement of numbers. We keep the term computation because it has de facto a broader meaning and absorbs almost every process that converges to a fixed point over time, although some might prefer the term information processing. By following our concept, computation is originally not bounded to analytical approaches and tasks but denotes the process of solving a problem. We state that the natural approach to computation is essentially different from the computational devices of the last 200 years.

The term computation in a very broad sense does not exclude our concept of embodied computation but it also does not include it explicitly. We define the concept of embodied computation:

Device of Embodied Computation – *A device performing embodied computation satisfies the following requirements:*

- *It consists of at least two levels with a micro-macro relationship, that is, a large number of embodied microscopic computational devices (MCD) of limited abilities form a single macroscopic computational device. This might be extended to a cascade of many levels.*

- *MCD are only loosely coupled among each other by means of short range communication.*
- *MCD are sensor, actuator, CPU, and memory in one entity.*
- *The system is highly adaptive (often also called self-organized).*
- *As a consequence of the above characteristics the solution to the macroscopically defined task emerges from the microscopic level and the whole system is extremely robust, that is, even a high percentage of damaged MCD has no influence to the effectiveness of the whole system.*

Such a device is able to solve problems that cannot be solved by one of its parts (MCD) or even by the sum of its parts. The solution to the problem emerges due to the high number of local interactions between the MCD. MCD could be, for example, robots of limited abilities, biological cells, or molecules.

The main differences of embodied computational devices compared to traditional concepts are:

- Memory and CPU are not separated.
- Due to the embodiment the physical world is a relevant part of computation.
- There is not a globally defined algorithm solving the task step by step.
- The solution is not repeatable and could be suboptimal or wrong.
- The system is robust rather than fragile.

We see two main benefits in investigating such computational systems: First, this kind of system might have the highest accomplishable robustness of all computational systems because it shows a maximum of local organization. Second, due to the involvement of embodiment and spatiality this might increase the possibilities of investigating and analyzing computation, for example, by applying methods of statistical mechanics or graph theory.

An example of an embodied system is a leaf with its stomatal apertures (Peak et al., 2004). The optimization problem needed to be solved by a leaf is to maximize CO_2 uptake and minimize water loss. This is done by the stomata interacting with neighboring stomata forming synchronized patches. Other examples are foraging ants (Bonabeau et al., 1999), a crowd of people trying to leave a room (Helbing et al., 2000), or a robotic swarm performing collective perception (Schmickl et al., 2007a).

A clear benefit is seen in the proposed approach to establish new ways of analyzing computation. This might be achieved by approved methods of statistical physics (Hamann and Wörn, 2007b; Schweitzer, 2003). Equations describing Brownian motion with drift can be applied, by interpreting the MCD as microscopic particles. Then the micro-macro link is solved by the methodology of physics. Especially, the analysis of the average behavior of a computational system, which is a hard task using traditional computer science, might be simplified.

An example scenario justifying the relevance of this hypothesis will be shown in Section 5.5 on page 113.

4.8 Discussion

4.8.1 Discussion of Limitations

Neglect of Velocities and the Markov Property

The most prominent limitation of this model is the assumption that the robots' motion is independent of their previous velocities and individual history. This independence becomes apparent in the Langevin equation (Equation 4.65 on page 63). The change of the position depends only on the current states of the robot's position $\mathbf{R}(t)$ and the functions $\mathbf{A}(\mathbf{R}(t),t)$, $B(\mathbf{R}(t),t)$, and $\mathbf{F}(t)$. First of all, $\mathbf{F}(t)$ introduces this independence of history because it is defined to be uncorrelated in time. Based on $\mathbf{F}(t)$ the diffusion term in the Fokker–Planck equation is derived, which is, thus, also only a correct approximation if many collisions occur. While it is safe to assume that $\mathbf{A}$ and B change slowly over time for the most relevant scenarios, they still do not incorporate the individual's history into the model. Thus, the underlying assumption of this model is the Markov property

$$\begin{aligned} &P(X_{t+1} = s(j_{t+1}) \mid X_t = s(j_t), X_{t-1} = s(j_{t-1}), \ldots, X_0 = x_0) \\ =&P(X_{t+1} = s(j_{t+1}) \mid X_t = s(j_t)), \end{aligned} \tag{4.72}$$

the probability of a transition to state $s(j_{t+1})$ depends only on the current state $s(j_t)$. The history of previous states ($s(t_k)$ with $k < t$) is irrelevant. In the presented model the state consists only on robot positions or robot densities respectively ($\mathbf{R}$ or ρ) and possibly on potential fields.

As we will show in Section 5 on page 77 this seems to be a valid assumption for a variety of swarm robotic applications. One can distinguish two classes of scenarios for which the neglect of velocities in connection with the Markov property is a useful assumption: First, the robot density is high, which results in many collisions making the history irrelevant. Second, the robot density is low but the robot's deterministic behavior has a homogenizing effect, that is, the robot density is balanced by deterministic effects and can, nevertheless, be approximately be modeled by diffusion.

The Markov property is certainly an approximation in this work – a fact that is also discussed by Gardiner (1985):

> [Markovian probabilistic equations derived from mechanics] are rarely truly Markovian – rather there is a certain characteristic memory time during which the previous history is important [...]. This means that there is really no such thing as a Markov process; rather, there may be systems whose memory time is so small that, on the time scale on which we carry out observations, it is fair to regard them as being well approximated by a Markov process. [...] Even if collisions of molecules are not accurately modelled by hard spheres, during the time taken for a collision, a finite change of velocity takes place and this will appear in the approximating Markov process as a discrete step. On this time scale, even the position may change discontinuously, thus giving the picture of Brownian motion as modelled by Einstein.

The neglect of velocities combined with the Markov property becomes an inappropriate assumption for other scenarios with low robot densities. For example, if there is a region X at which the robots' velocity distribution is inhomogeneous (e.g., one preferred direction) and they move straight into a region Y with a low probability of collisions then we have a dependence on the robots' individual history. The direction a robot adopted in region X is propagated over time an space into region Y. Thus, we need to know the robot's individual velocity and/or history to predict its next position. As an extension to the proposed model this history could be incorporated in the model state allowing the Markov property again. This could, for example, be done by including probability densities for the direction in each point (e.g., a mesoscopic model similar to the Boltzmann equation as discussed in Section 2.1.3 on page 6.

Imbalance between the Langevin and the Fokker–Planck Equation

In using the model we abstract from explicit agent-agent interactions. In case of the Fokker–Planck equation this is done by applying, for example, sigmoid functions that are based on the robot density. Such functions are used to model deterministic (e.g., turn around, if you see less than five robots) or random movements (e.g., increase of collision avoidance behavior with increasing density). The meaning of the robot density ρ in the functions $\mathbf{A}$ and B is unclear in case of the Langevin equation. On the micro-level this abstraction does not have to be done because the positions of the robots are explicit and known. However, this might cause high computational complexity and the one-to-one correspondence between the Langevin and the Fokker–Planck equation would break. Alternatively, an approach based on densities is used. For example, the number of neighboring robots can be estimated by counting the robots within a smaller radius (implying less computational cost). Then the expected number for the full communication radius is estimated, see Figure 4.10. This is an agent-based approach to the robot density.

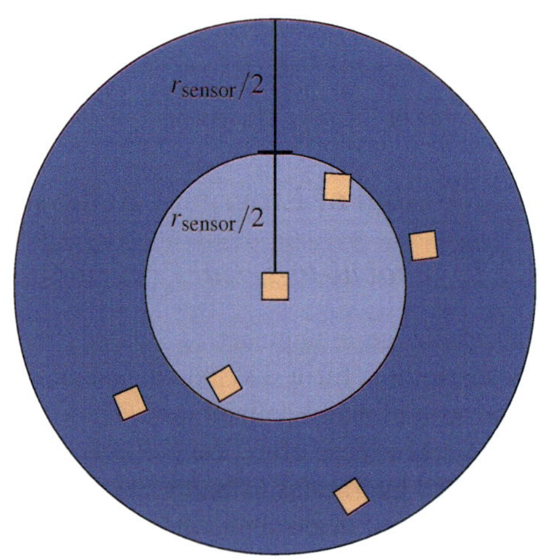

Fig. 4.10 Agent-based approach to the robot density. Neighboring robots are only counted within a distance of $r_{sensor}/2$ (two). Thus, the estimated robot density for the full sensor range is $\frac{6}{\pi r_{sensor}^2}$.

The next abstraction step is to use the explicit or the above estimation method only for an initial phase during which the robot densities in dependence on space are measured. Based on these measurements an estimated probability density is determined which is used and improved thereafter (i.e., establishing a histogram for the robot densities, see Figure 4.11). This is a space-based approach to the robot density. An effect is that robots can be simulated as stand-alone once the density is sufficiently well approximated.

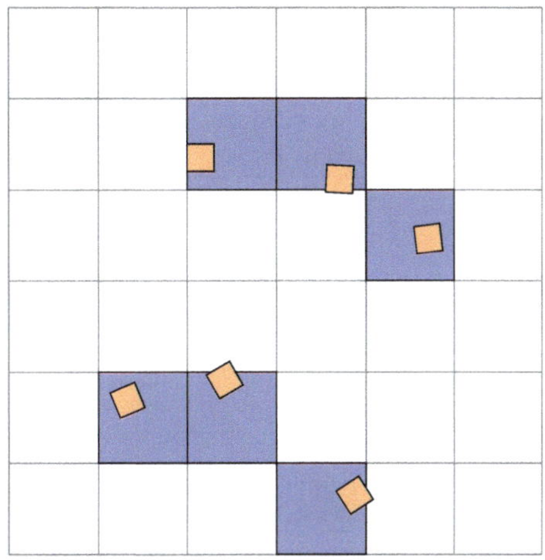

Fig. 4.11 Space-based approach to the robot density, estimation using a histogram

Although the above presented methods allow the usage of the robot density ρ in the functions $\mathbf{A}$ and B also for the Langevin equation, an imbalance between both equations persists. The origin of this problem is the possibly fundamental difference between approaches based on ensembles and trajectories, see also Section 3.1.4 on page 29.

Lack of Concrete General Methodological Principles

The creation and the application of models is a big piece of science itself. The chance for a fully automatic model development tool seems, therefore, to be improbable. A detailed and generic step-by-step manual for the development seems to be problematic as well. Even for the relatively bounded field of swarm robotic application scenarios the strive for such a description might be doomed to fail. This challenge reflects the possibly essential creative input needed to understand emergence as discussed in Section 3.2.1 on page 35.

4.8.2 Discussion of Benefits

The main benefit of the proposed model is the strong connection between the micro-level (trajectories, Langevin eq.) and the macro-level (ensembles, Fokker–Planck eq.) by means of the analytic derivation as shown in Section 4.1.3 on page 47. This micro-macro connection is visualized once more in Figure 4.12. By bridging the gap between the micro- and the macro-level the model is superior to phenomenological models concerning explanatory power. For example, a diffusion term in a phenomenological model has the only meaning that the observed macro-behavior can be described by such a term. However, it does not explain how or if this macro-behavior might be achieved as an effect of micro-behavior. By defining the $\mathbf{A}$- and the B-function two models are defined: a microscopic and a macroscopic model. Hence, the presented model might also serve as a tool to find a better understanding of emergence.

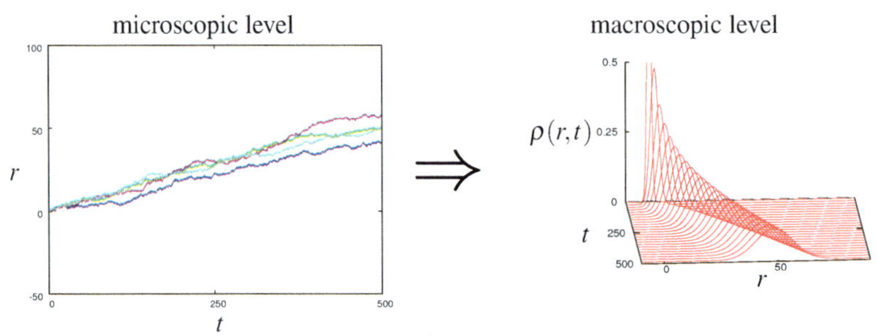

$\dot{\mathbf{R}}(t) = -\mathbf{A}(\mathbf{R}(t),t) + B(\mathbf{R}(t),t)\mathbf{F}(t)$ $\quad$ $\frac{\partial \rho(\mathbf{r},t)}{\partial t} = -\nabla(\mathbf{A}(\mathbf{r},t)\rho(\mathbf{r},t)) + \frac{1}{2}Q\nabla^2(B^2(\mathbf{r},t)\rho(\mathbf{r},t))$

Langevin eq. Fokker–Planck eq.

Fig. 4.12 The connection between micro-level (trajectories, Langevin eq.) and macro-level (ensembles, Fokker–Planck eq.).

The general formulation of the model offers the possibility to apply it generically to a potentially vast variety of application scenarios, which is also shown in Chapter 5 on page 77. The equations of robot motion can be coupled with any kind of equations describing physical phenomena in the environment. Therefore, the model can be coupled with models of diffusion processes, fluid dynamics, electromagnetic waves, gas kinetics, etc.

As already discussed in Section 4.2.2 on page 58 the proposed model has a very concise description because of its symbolic character. This short description allows an easy comparison of different swarm behaviors and to other models. This degree of conciseness and comparability is hardly achievable using microscopic descriptions such as simulations.

4.8.3 Discussion of the Relevance to Computation

Why should we investigate alternative computational systems such as the proposed concept of embodied computation? The traditional computational models such as the Turing machine or the von Neumann architecture seem to cover only a small part of possible forms of computation. Although these models have helped immensely from the past until today to build fast computers, to analyze, and design sophisticated algorithms, there might be complete facets yet uncovered because we have focused too much on similar models.

For example, consider the restriction of separating memory and CPU. A typical characteristic of natural, computational entities is the unity of all components: "In the case of leaves, stomata are simultaneously the sensors of external information, the processing units that calculate how gas exchange regulation should occur, *and* the mechanisms for executing the regulation." (Peak et al., 2004) Restating this, MCD are sensor, computing unit, memory, and actuator in one.

Additionally, the memory of an embodied computational system consists not only of the sum of the distributed memories but of three parts: First, there is the internal memory of the MCD, if they have some at all. Second, the environment serves as a memory. For example, this can be done by stigmergy: A robot changes the environment locally. Later another robot senses this alteration and reacts accordingly to this information. Third, the positions of the mobile MCD can be sensed by others and can serve as memory.

The essential implication of embodiment is that many interactions between the agents are not executed following an artificial protocol but the natural laws of physics. Thus, both the parameters and the variables are determined and "administrated" by the physical world. In contrast to the explicit representation of numbers hold in an electronic memory, being encoded, sensitive to slight errors, and treated by programs according to mathematical laws.

There is no global algorithm but it rather emerges from the local algorithms defining only the behavior of the MCD. The global behavior that actually defines the information processing is only implicitly determined. An embodied computational system does not calculate like adding or subtracting numbers and it does not solve problems exactly.

Regarding robustness, it seems to be hardly possible, in general, to have a maximum amount of robustness within a single computational system and assuring to find the exact solution at the same time. In fact this seems to be a trade-off as the common trade-off between exploration and exploitation (Holland, 1992). If the system computes an exact solution then there is a certain task assigned to each component creating the possibility of a single point of failure. However, this consideration is only true if the robustness of a macroscopic computational system consisting of many microscopic parts is considered. There is a difference, if we have redundancy on the macroscopic level: Only if we have multiple macroscopic computational systems we can have robustness and exactitude at the same time. On this higher level, there is no difference in the robustness between a bunch of personal computers and a bunch of embodied computational systems.

4.9 Summary

In this chapter we have motivated and presented our model. Models serve as a support during the algorithm design phase in swarm robotics. This is done in an iterative way by alternately adapting the model to the algorithm, and the algorithm to the predicted global behavior. By using the model to predict the swarm behavior the human intuition is bypassed, which is overburdened by the high number of agent–agent interactions in the systems.

Our model builds up on the model of Brownian motion which is extended to a generic model of communicating robot swarms. Important is the connection between the micro- and the macro-level. This connection leads to two corresponding models. The Langevin equation describes the swarm behavior in trajectories and the Fokker–Planck equation is a description based on ensembles. We have discussed the challenge of modeling communication analytically and have pointed out alternative approaches. General methodological principles were given highlighting the possibilities of generalization. Indicating the option of regarding our model as a computational model we have presented the concept of embodied computation. This concept allows to interpret the swarm dynamics as a form of computation. We have concluded the chapter by a detailed discussion of the pros and cons of the proposed concepts.

Chapter 5
Validation by Results of Experiments and Simulations

Models are, for the most part, caricatures of reality,
but if they are good, then, like good caricatures,
they portray, though perhaps in distorted manner,
some of the features of the real world.
The main role of models is not so much to explain and to predict
- though ultimately these are the main functions of science -
as to polarize thinking and to pose sharp questions.

Frank B. Knight in "Essentials of Brownian Motion and Diffusion"

We validate our model and the computational interpretation of swarms with algorithms implementing collision-based aggregation, collective perception, emergent taxis, foraging, and "random-tree" aggregation.

Several case studies were carried out to approve the utility and to validate the accurateness of the proposed model. One of these studies is based on experiments with real robots the other studies were performed using simulators. Each of the following examples is a little research project of its own although the model framework is a great support. There is still a respectable amount of creativity and insight necessary to obtain a satisfying result. The aim of these case studies is not to compare the swarm algorithms to alternatives but rather to take them for granted and to check what can be predicted and with what accuracy.

5.1 Collision-Based Adaptive Swarm Aggregation

The swarm algorithm, that is investigated in this case study, was designed and experimentally evaluated by Schmickl et al. (2008). For the experiments 15 Jasmine swarm robots were used, which have been developed at the Institute for Process

H. Hamann: Space-Time Cont. Models of Swarm Rob. Sys., COSMOS 9, pp. 77–122.
springerlink.com

Control and Robotics here at the Universität Karlsruhe (TH) in cooperation with the University of Stuttgart (Kornienko et al., 2005a; Szymanski and Wörn, 2007). In this scenario, these robots were controlled by a bio-inspired algorithm, which is derived from honeybees' navigation behavior in a temperature gradient. The following specific modeling approach and the results were previously reported in a short form in Hamann and Wörn (2008b), Hamann et al. (2008), and Schmickl et al. (2009). Further studies with focus on the symmetry breaking capabilities of this system are reported in Hamann et al. (2010).

5.1.1 The Swarm Robot "Jasmine"

The swarm robot Jasmine (see Figure 5.1 and Jasmine (2008)) was developed especially for swarm robot research. Despite its small size of about $30 \times 30 \times 30\ \text{mm}^3$, it has good local communication abilities and a far distance scanning and distance measuring sensor. The good communication abilities result from six infrared sensors and emitters arranged around the robot with a displacement of 60 degrees. These sensors are also used for short distance measurements. The far distance measuring sensor is hooked to the front of the robot. Two differentially driven wheels give this robot a high manoeuvrability at a high speed. Generation three of the robot, that was used here, has an Atmel Mega 168 micro-controller with 1 Kbyte RAM and 16 Kbyte Flash. A single LiPo battery pack is sufficient for up to two hours of motion and optical encoders are available that allow odometric measurements in the mm-range.

Different from some other swarm robots, Jasmine supports only local communication. Long distance communication via radio frequency is not implemented and does not correspond with the views of the construction team about swarm robot capabilities (see also "The minimalist approach", Section 2.3.3 on page 14). In addition, it is possible to attach an optional sensor board with two photodiodes on top of the robot.

Because of its design for low cost and the artificial restriction to use no long distance communication the Jasmine robot might seem to be below the

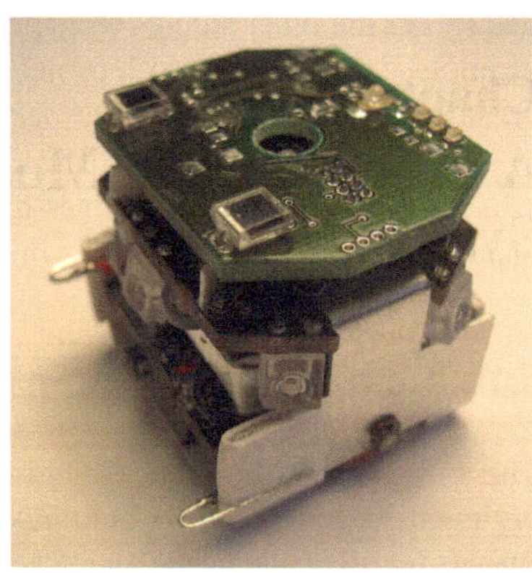

Fig. 5.1 The Jasmine swarm robot

technological state-of-the-art. However, this makes it a suitable testbed for even much smaller robots because besides the locomotion there are no conceptual problems for miniaturization (see Section 2.3.2 on page 12).

5.1.2 Scenario

The underlying control algorithm is called *BEECLUST* (Schmickl et al., 2008). To transfer the honeybees' behavior to a robotic swarm, the spatial temperature distribution is replaced by light emitted by lamps mounted above the arena. In the following experiments, the collective decision making ability of the swarm is investigated by testing, whether the robots are able to aggregate predominantly at the brightest spots. The robots move within an arena of 150 cm $\times$ 100 cm, which is lighted by one or two lamps at the sides. The luminance of these lamps can be varied between two configurations: dimmed (about 390 lux at the center of the light spot) and bright (about 1100 lux).

The robots' behavior is defined as follows (BEECLUST algorithm):

1. All robots move (ideally) in straight lines.
2. If a robot detects an object in front by analyzing reflections of emitted infrared (IR) pulses it stops and listens for IR signals without emitting pulses itself.
3. If there are no relevant IR signals, the robot assumes the object to be an obstacle (e.g., the arena wall), turns randomly and continues with step 1.
4. If the robot detects foreign IR signals, it assumes the object in front to be another robot. The stopped robot then measures the local luminance. The higher the luminance, the longer it waits at that place.
5. After the waiting time has elapsed, the robot turns 180 degrees and continues with step 1.

The key component in this swarm robotic system is the dependency of the waiting time of a clustered robot on the locally measured luminance. The used sensor scales approximately linearly from 0 to 180 for corresponding luminance between 0 lux and about 1500 lux. We call this sensor value S. The following equation is implemented in the control algorithm of the robots to map these sensor values S to waiting times

$$w(S) = \frac{w_{\max} S^2}{S^2 + 7000}, \tag{5.1}$$

for a constant $w_{\max}$ (see Table 5.1 on page 83 for the parameter values that were used here).

The resulting sigmoid function for the direct mapping of light intensities to waiting times is shown in Figure 5.2 on the next page.

The experiments carried out by Schmickl et al. (2008) showed: This simple algorithm leads to collective aggregation of robots below the light source. The intensity of the light influences the total number of robots that aggregate below the light source. Besides that, also the spatial pattern of aggregated robots at the light source

Fig. 5.2 The waiting time a robot waits after an encounter of another robot is non-linearly correlated with the measured local luminance. Robots measure this local luminance only after encounters with other robots. This curve is defined by Equation 5.1 on the page before, which is implemented in the robots' control algorithm, and the (linear) behavior of the sensor (Schmickl et al., 2008).

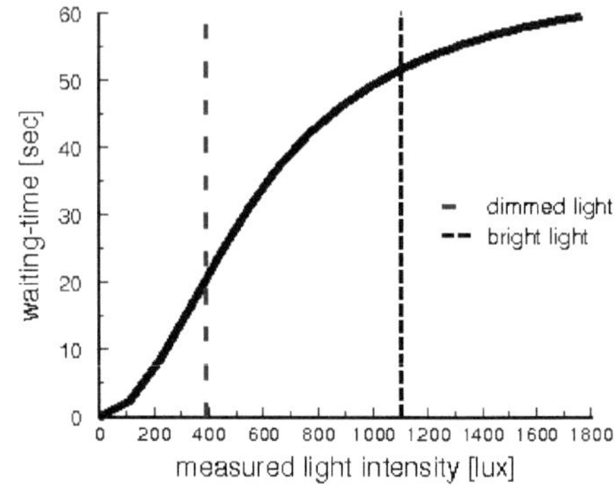

changes with different intensities of the light: With dimmed light, fewer robots aggregate at the light source, but these robots get closer to the center of the light spot. With brighter light intensity, more robots aggregate, but they stop farther away from the center of the spot (see Figure 5.3).

(a) Dimmed light. (b) Bright light.

Fig. 5.3 Frames taken by the tracking video camera mounted above the arena during the robot experiments, showing typical results: In bright light conditions, more robots aggregate below the lamp (right side of arena) than in dimmed light conditions. With bright light, more robots form (often ring-shaped) clusters around the light spot. With dimmed light, robot clusters are formed closer to the center of the light spot (Hamann et al., 2008).

In the following, we focus on scenarios with different light configurations. Alternating we will have one local and one global light maximum (emitted by two lamps) or just a single maximum. In the experiments, the robots preferentially aggregated at the spot with the higher luminance. In the first scenario, the light intensities were changed (in four phases of 180 seconds each, see sun-like symbols in Figure 5.5 on page 85) at several points in time. The robotic swarm quickly adapted its prior decision and reallocated robots between the two clusters to reach an adaptive and robust solution. In the second scenario we just analyze the swarm's reaction within 180 seconds to turning on a single dimmed or bright light.

5.1.3 Model

Before we present a model of the actual robot behavior we model the environment, that is, the light distribution. The light intensity (illuminance) decays quadratically in the distance to the light source. Furthermore, the reflector of the lamp is modeled as a second light source at the reflector's focal point below the actual lamp. We get a spatiotemporal description of the luminance distribution

$$\begin{aligned} e(\mathbf{r},t) =& c_1(t)(h^2+\text{dist}(\mathbf{r},\mathbf{L_1})^2)^{-1}+c_1(t)((h-f)^2+\text{dist}(\mathbf{r},\mathbf{L_1})^2)^{-1} \\ &+c_2(t)(h^2+\text{dist}(\mathbf{r},\mathbf{L_2})^2)^{-1}+c_2(t)((h-f)^2+\text{dist}(\mathbf{r},\mathbf{L_2})^2)^{-1}[cd/m^2], \end{aligned} \tag{5.2}$$

for positions $\mathbf{r}$ in the arena, time t, two lamps at a height h [m] with position (projected to the arena) $\mathbf{L_1}$ and $\mathbf{L_2}$, focal points f [m] below the lamps, time-variant intensities $c_1(t)$ [cd] and $c_2(t)$ [cd] in candela, and a function $\text{dist}(\mathbf{a},\mathbf{b})$ [m] giving the distance between points $\mathbf{a}$ and $\mathbf{b}$. The sensor values associated with positions $\mathbf{r}$ in the arena are modeled linearly as discussed above and the maximal value delivered by the sensor is 180:

$$s(\mathbf{r},t) = \min\left(180, \frac{64}{375}e(\mathbf{r},t)\right). \tag{5.3}$$

The waiting times (based on Equation 5.1 on page 79) associated with positions $\mathbf{r}$ are defined by

$$w(\mathbf{r},t) = \frac{w_{\max}s^2(\mathbf{r},t)}{s^2(\mathbf{r},t)^2+7000}. \tag{5.4}$$

Now the environment is defined and we move on to the model of the swarm behavior. We distinguish two states: *moving* robots ρ_m and *aggregated* robots ρ_a. Hence, we get two PDE and follow the formalism of Equation 4.71 on page 67 (system of Fokker–Planck equations with transition probabilities). We consider the density of moving robots $\rho_m(\mathbf{r},t)$ at position $\mathbf{r}$ and at time t. On the one hand, ρ_m is reduced by a number of robots which switch to the *aggregated* state. Say, these transitions depend on a transition rate $t_{m,a}$, which we do not specify for now. On the other hand, ρ_m is increased by beforehand aggregated robots ρ_a that switch back to the *moving* state. Say, these transitions depend on a transition rate $t_{a,m}$. Concerning the motion model we can simplify the Fokker–Planck equations here due to the simple control algorithm. Although the principle motion of the robots is in straight lines, we assume that we can sufficiently approximate it by a diffusion term, that is, we drop the drift term for ρ_m. This simplification is discussed below (see Section 5.1.5 on page 87). Furthermore, the aggregated robots do not move, which simplifies the equation for ρ_a essentially. We get

$$\frac{\partial \rho_m(\mathbf{r},t)}{\partial t} = D\nabla^2 \rho_m(\mathbf{r},t) + t_{a,m}(\mathbf{r},t)\rho_a - t_{m,a}(\mathbf{r},t)\rho_m(\mathbf{r},t) \tag{5.5}$$

$$\frac{\partial \rho_a(\mathbf{r},t)}{\partial t} = t_{m,a}(\mathbf{r},t)\rho_m(\mathbf{r},t) - t_{a,m}(\mathbf{r},t)\rho_a(\mathbf{r},t), \tag{5.6}$$

for a diffusion coefficient D. This system of equations can be reduced to a single equation. This is done by specifying $t_{m,a}$ and $t_{a,m}$. Let us provisionally ignore that the waiting times are dynamic. As said above, ρ_m is increased by beforehand aggregated robots that switch back to the *moving* state. These beforehand stopped robots are that fraction of robots, whose waiting time $w(\mathbf{r})$ has elapsed at time t. This fraction is equal to the fraction of robots which had stopped at time $t - w(\mathbf{r})$. Thus, we note that

$$t_{a,m}(\mathbf{r},t)\rho_a(\mathbf{r},t) = t_{m,a}(\mathbf{r},t - w(\mathbf{r}))\rho_m(\mathbf{r},t - w(\mathbf{r})). \tag{5.7}$$

Thus, a formalism of just one transition rate is possible. We do this by defining the stopping rate $\varphi(\mathbf{r},t) = t_{m,a}(\mathbf{r},t)$, which is space- and time-dependent and describes the ratio of stopping robots.

The stopping rate φ is derived by using simple collision theory (Trautz, 1916) as presented in Section 3.1.5 on page 30. At first, we compute the collision density $Z_{f,a}$ (collisions per time and area) for moving robots colliding with aggregated robots. The relevant area as introduced above is $C_{f,a} = 2\sigma v$ (Equation 3.1 on page 31). Here, we take for the cross section the sensor range $\sigma = r_{\text{sensor}}$. Area $C_{f,a}$ is populated by $C_{f,a}\rho_a$ aggregated robots giving also the number of expected collisions for a single moving robot. Multiplying by ρ_m gives the collision density as defined by collision theory

$$Z_{f,a}(\mathbf{r},t) = C_{f,a}\rho_a(\mathbf{r},t)\rho_m(\mathbf{r},t). \tag{5.8}$$

The collision density of moving robots colliding with moving robots is derived by similar considerations except for two differences. First, we have to use the mean relative speed leading to an area of $C_{f,f} = r_{\text{sensor}}\frac{4v}{\pi}$ (Equation 3.2 on page 31). Second, the result needs to be divided by two as each collision is counted twice. We get

$$Z_{f,f}(\mathbf{r},t) = \frac{1}{2}C_{f,f}\rho_m^2(\mathbf{r},t). \tag{5.9}$$

The stopping rate is the fraction of moving robots, that collide with another robot and detect this collision as well. The probability of detecting a collision successfully is given by P_{detect} which is multiplied to the sum of both collision densities. This probability depends on the software implementation, the sensor arrangement, and on the quality of the sensors. Here, it is set to $P_{\text{detect}} = 0.5$, which corresponds to the observations during the experiments. $Z_{f,f}$ is multiplied by two because each moving-to-moving collision can potentially convert two moving robots into two aggregated robots (this reverts the division by two in the derivation of Equation 5.9). We get

$$\varphi(\mathbf{r},t) = \frac{P_{\text{detect}}}{\rho_m(\mathbf{r},t)}(Z_{f,a}(\mathbf{r},t) + 2Z_{f,f}(\mathbf{r},t)). \tag{5.10}$$

Furthermore, we have to introduce a sigmoid function $\Gamma(\mathbf{r},t) \in [0,1]$ limiting the maximum robot density. It is multiplied to diffusion D such that it can be interpreted as a space- and time-dependent diffusion. The effect is a slowdown of the robot density flow in regions of high density. We define

$$\Gamma(\mathbf{r},t) = (1+\exp(\gamma_1(\rho_m(\mathbf{r},t)+\rho_a(\mathbf{r},t))/\rho_c - \gamma_2))^{-1}, \tag{5.11}$$

for constants γ_1, γ_2, and a "critical" density ρ_c at which the robots' movements become almost impossible. We choose $\rho_c = 1/(\pi r_{\text{sensor}}^2)$. The shape of Γ, the value of ρ_c, and the diffusion coefficient D are free parameters that are fitted to the scenario without much overhead. Now the model is defined by a single equation

$$\begin{aligned} \frac{\partial \rho_m(\mathbf{r},t)}{\partial t} =& D\Gamma(\mathbf{r},t)\nabla^2\rho_m(\mathbf{r},t) - \rho_m(\mathbf{r},t)\varphi(\mathbf{r},t) \\ &+ \rho_m(\mathbf{r},t-w(\mathbf{r}))\varphi(\mathbf{r},t-w(\mathbf{r})). \end{aligned} \tag{5.12}$$

Table 5.1 Parameters used in the model

Parameter	value
arena dimension	150cm × 100cm
avg. velocity v	30cm/s
sensor range r_{sensor}	7.5cm
P_{detect}	0.5
$w_{\max}$	66
diffusion D	1×10^4
critical density ρ_c	$1/(\pi r_{\text{sensor}}^2)$
γ_1	20
γ_2	13

In the scenario under investigation here, the waiting time $w(\mathbf{r},t)$ is dynamic. Thus, the waiting times of robots having stopped at several different times in the past might elapse simultaneously. We have to sum the robot fractions over all these times to model the transition from aggregated to moving robots correctly. Hence, the last summand of Equation 5.12 becomes

$$\sum_{w \text{ with } w(\mathbf{r},t')+t'=t} \varphi(\mathbf{r},t-w)\rho_m(\mathbf{r},t-w), \tag{5.13}$$

which is the sum over all waiting times $w(\mathbf{r},t')$ that elapse at time t. Hence, the condition for the sum is $w(\mathbf{r},t')+t'=t$.

The boundary conditions are set to total isolation (no robots leave or enter). The initial condition is a homogeneous distribution of robots in the dark half of the arena. All of them are moving robots; there are no stopped robots at the beginning. We get an initial value problem which was solved numerically.

Most of the occurring parameters can be determined well, such as the spatial luminance distribution and the mapping of this luminance to sensor values. The shape of Γ, the value of ρ_c, and the diffusion coefficient D are free parameters that are fitted in a simple process to the scenario. Diffusion D is used to adjust the homogenization speed of the moving robot density $\rho_m(\mathbf{r},t)$. This fitting of free parameters is only a quantitative adaptation. The main qualitative features of the model are determined by its form and the other parameters.

5.1.4 Results

In Figure 5.4 the waiting times for all positions in the arena are given as defined by Equations 5.1 on page 79, 5.2, and 5.3 on page 81. The four diagrams correspond to the four light configuration phases (right dimmed, right dimmed and left bright, right bright and left dimmed, left dimmed). Plateaus around $w = 60$ are identified For the bright-light-configurations. These are caused by the maximal sensor value of $s(\mathbf{r},t) = 180$.

In Figure 5.5 on the next page we compare the predicted number of aggregated robots at the left light, at the right light, and all aggregated robots to the data of the experiments by Schmickl et al. (2009) over time. The model was manually fitted using the free parameters of Γ and D. The resulting correspondence is good.

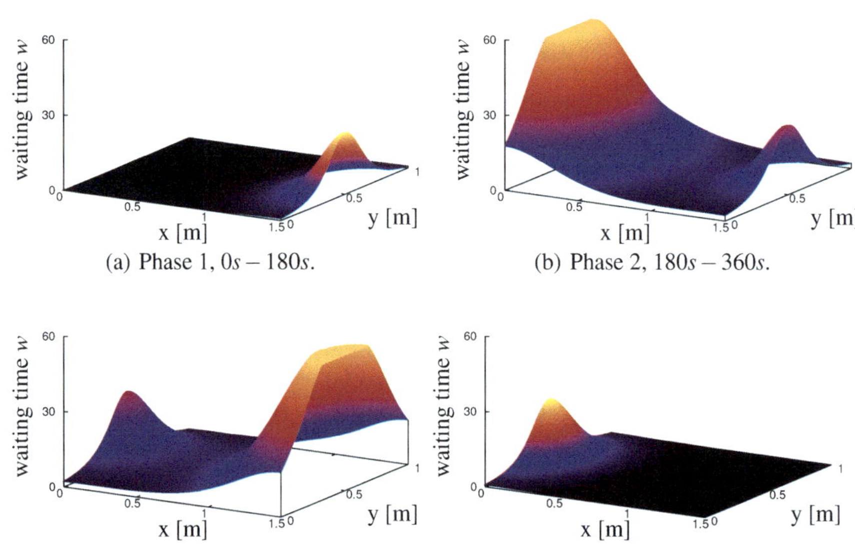

(a) Phase 1, $0s - 180s$.

(b) Phase 2, $180s - 360s$.

(c) Phase 3, $360s - 540s$.

(d) Phase 4, $540s - 720s$.

Fig. 5.4 Sensory map showing the sensor values as defined by Equations 5.1 to 5.3 on pages 79–81 for the four phases of the experiment (Schmickl et al., 2009)

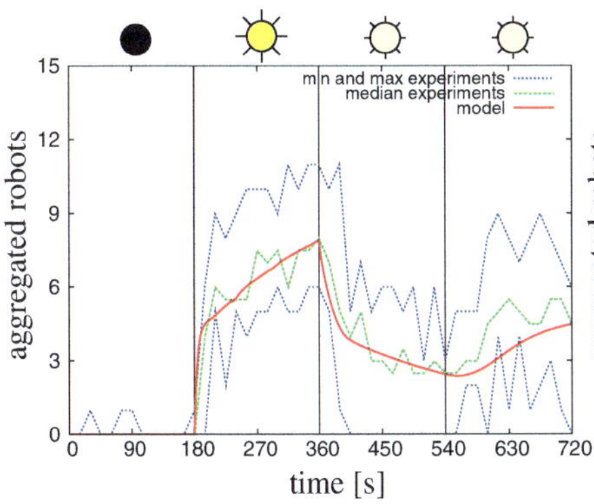

(a) Aggregated robots at left light.

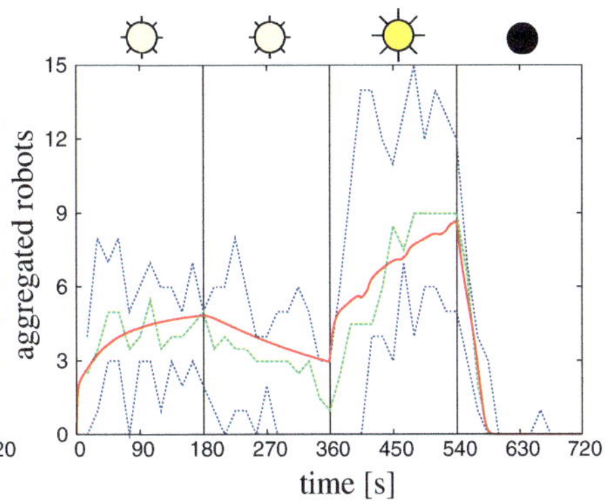

(b) Aggregated robots at right light.

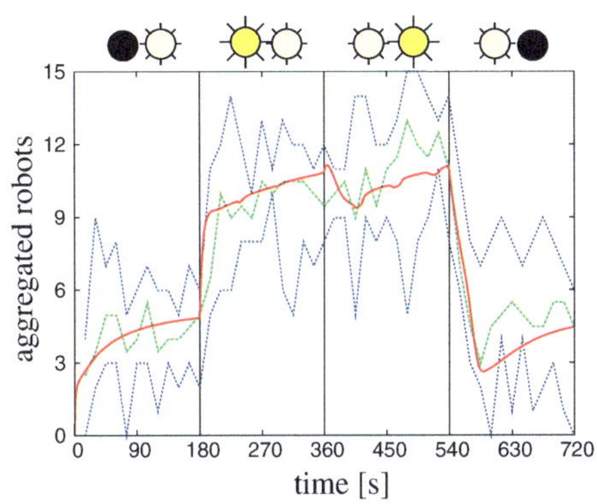

(c) Sum of aggregated robots at both lights.

Fig. 5.5 Comparison of the experiments with real robots (Schmickl et al., 2009) to the model. The sun-like symbols indicate the light intensity emitted by the corresponding lamp: A black sun indicates that the lamp was turned off in this phase, a light yellow sun indicates a dimmed light, and a bright yellow sun indicates a bright lamp (data of the experiments is based on six repetitions).

In Figure 5.6 on the following page we compare a frame of the tracking camera which was recorded during the second phase of the experiment to the density prediction of aggregated robots by the model. This cannot serve as a validation but sampling statistically significant, highly resolved robot densities from experiments would generate an immense overhead that has not been invested yet. At least the tendency of the robots aggregating around the central spot of the bright light instead of aggregating right at the brightest spot can be seen in the photo as well as in the model prediction.

In an additional experiment another light configuration was tested. At time $t = 0$ one bright light or one dimmed light was turned on and the reaction of the swarm

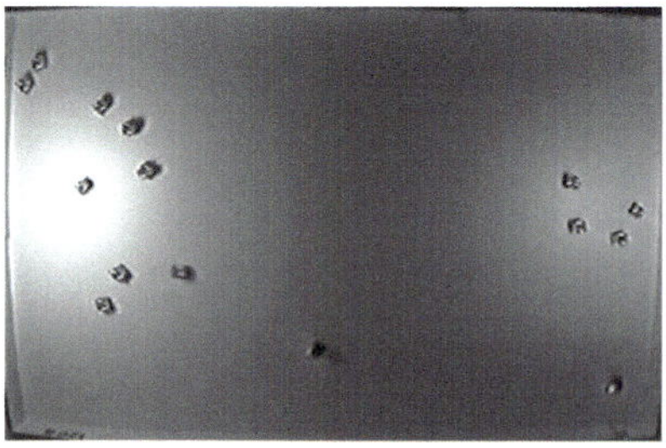

(a) Photo of the arena showing a typical configuration.

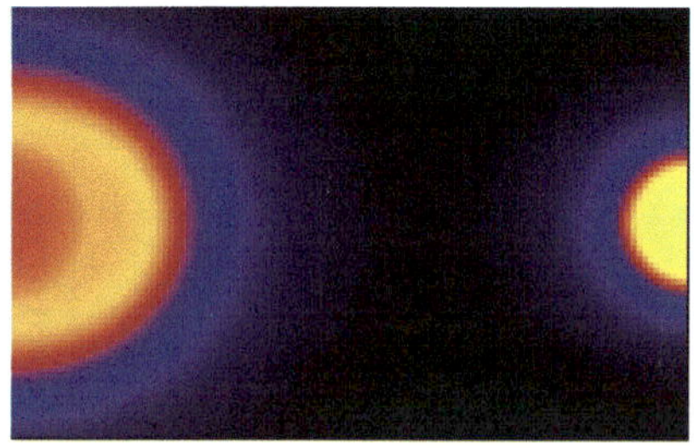

(b) Expected density of aggregated robots (higher densities are brighter).

Fig. 5.6 Comparing a swarm configuration observed during the experiments in phase two to the prediction of the model. The brighter light is on the left side and the dimmed light on the right.

was measured for 180 seconds. This corresponds to the first phase of the previous scenario where one dimmed light was turned on.

The arena was partitioned into five logical zones to analyze the spatial distribution of robots (see Figure 5.7). Zone 1 includes the brightest spot of the lamp and has a radius of $R_1 = 11cm$. Zones 2, 3, and 4 are ring-shaped with an outer radius of $R_2 = 22cm$, $R_3 = 33cm$, and $R_4 = 66cm$ respectively. The fifth zone covers the remaining area.

Fig. 5.7 Partition of the arena in zones (Hamann et al., 2008)

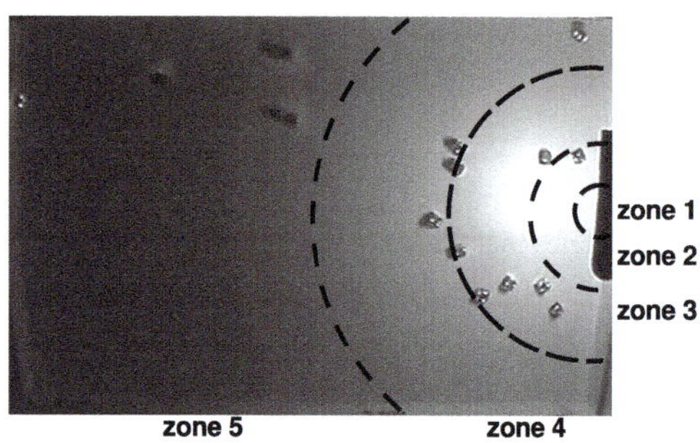

In Figure 5.8 on the facing page these two scenarios are compared by focusing on the temporal evolution of aggregated robots at the two central zones (1 and 2) and the two peripheral zones (3 and 4). There are discrepancies for $t < 90$ while the accuracy for $t > 90$ is good.

In Figure 5.9 on page 88 the average over the last 60 seconds ($120s - 180s$) of aggregated robots in zone 1, 2, 3, and 4 of the experiments and the model are compared. An inaccuracy of the model is found for zone 1 for both cases, the dimmed and the bright light. There, the aggregated robots are significantly overestimated.

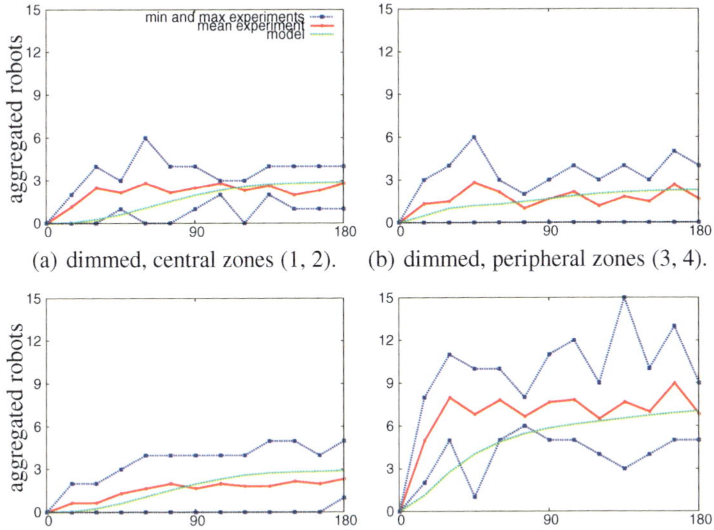

(a) dimmed, central zones (1, 2). (b) dimmed, peripheral zones (3, 4).

(c) bright, central zones (1, 2). (d) bright, peripheral zones (3, 4).

Fig. 5.8 Comparing the numbers of aggregated robots over time at the central area (zones 1 and 2) and the peripheral area (zones 3 and 4) as predicted by the model to mean, min., and max. of six robot experiments (Hamann et al., 2008).

5.1.5 Discussion

The most significant inaccuracies are found in the last phase at the left light and in the second and third phase at the right light (cf. Figure 5.5 on page 85). Similar problems arise in the other experiment in case of the dimmed light at zones 1 and 2 (cf. Figure 5.8(a)) and in case of the bright light at zones 3 and 4 (cf. Figure 5.8(d)). This seems to be a problem induced by the assumption of a diffusion process. The robots in the experiment show rather a so-called correlated random walk, since they move straight between relatively seldom occurring collisions. Having said that, the deterministic component of the robot behavior (moving in straight lines) does actually lead to a homogenized robot density just as a diffusion process. A macroscopic model of this time-correlated behavior is barely possible. In the absence of collisions, a robot moves straight, that is, it propagates its own directional information through space and time. This information is not statically bound to the environment because it just appears at one spot in form of the robot and leaves with the robot. This violates the underlying Markov property of our model since the history of a robot matters. It is very likely that this can only be solved by using a model which is a bit closer to the microscopic level. This could be a mesoscopic model in correspondence to the Boltzmann equation in physics (as discussed in Section 2.1.3 on page 6). The accuracy of the space-continuous model would be increased with higher numbers of robots in the experiment, that is, the robot density and, thus, the collision frequency would be increased. For the presented scenario with low density, the model was adapted by using

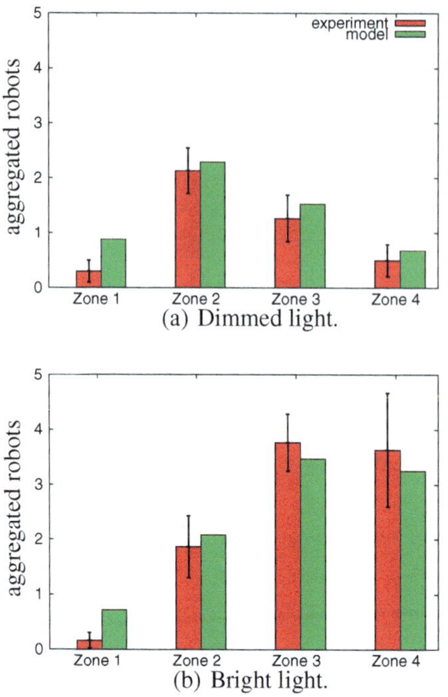

Fig. 5.9 Comparing the numbers of aggregated robots in each of the four considered zones as predicted by the two models to the mean numbers of aggregated robots in each zone during the last 60 seconds of 6 robot experiments; error bars indicate the upper bound of the 95% confidence interval (Hamann et al., 2008).

a high diffusion coefficient D. This leads to a faster "mixing" or homogenization of the density throughout the arena. This is a counter-intuitive approach as it actually would be interpreted as a higher number of collisions in the image of a diffusion process. However, this is the only way of overcoming the lack of temporal correlations and emulating the observed behavior.

Another approach could be the application of a Lévy walk in an inhomogeneous medium as reported by Barthelemy et al. (2008). The step length of a Lévy walk has power-law distribution leading to the occurrence of very long jumps. This results in super-diffusion, that is, the average displacement increases super-linear in time.

The other inaccuracy of the model is the overestimation of aggregated robots in zone 1 (cf. Figure 5.9). We assume this problem is caused by the U-turn behavior described in step 5 of the robot algorithm which is not modeled. Due to this U-turn, the probability of moving away from the light spot center is higher for a woken up robot than moving towards it. Additionally, even if there is only a rather low density, for example, in zone 2 these few robots might already block the way to zone 1. Again, there seems to be no possibility of incorporating this behavior induced by directional information and history in a purely macroscopic model of reasonable complexity. Again, an equivalent to the Boltzmann equation could be helpful and would lead, thus, to a mesoscopic approach.

5.2 Collective Perception

The case study reported in this section was previously published in Hamann and Wörn (2007b) and Hamann and Wörn (2008b). Collective perception is a process in which a group of autonomous agents collectively extracts the main features of an object or, more general, of the environment. These features can not be observed in principle or at least not within a reasonable time frame by a single agent. An algorithm inspired by honey bees implementing collective perception on a relatively low level of complexity was presented by Schmickl et al. (2007a). Their approach is called *trophallaxis strategy* in appreciation of the underlying behavior in social insects, for example, mouth-to-mouth transfer of food in bees.

We give a short description of the scenario in the following section. The interested reader is referred to (Schmickl and Crailsheim, 2008, 2006; Schmickl et al., 2007c,b,a) for more information. Subsequently, we explain the trophallaxis strategy and give a derivation of the model.

5.2.1 Scenario

The robots move in a bounded planar area. There are two designated areas of different sizes that we call "target areas" in the following (cf. Figure 5.10). The task of the swarm is to aggregate at the bigger target area. However, measuring the dimension of an area is beyond a single robot's abilities. Using the trophallaxis strategy even irregular areas, for example, scattered and loosely connected areas, can be collectively perceived, although we restrict our investigations to simple circular shapes here. Many applications are possible using this behavior as a component, such as cleaning, monitoring, or transportation scenarios.

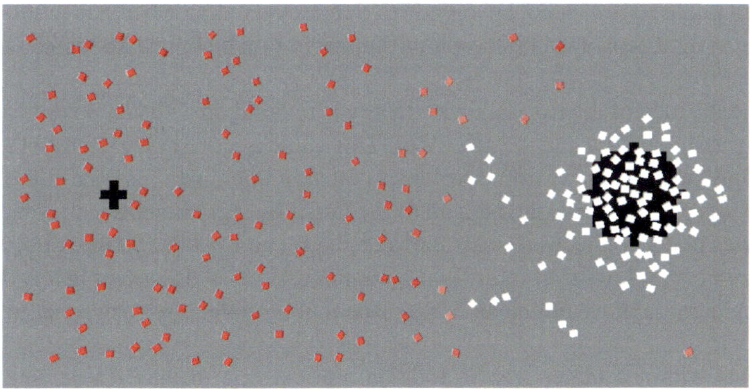

Fig. 5.10 Screen-shot of the simulation showing two circular target areas of different size and cubic robots with their value of the potential field indicated by their brightness, the brighter the higher the value (Hamann and Wörn, 2008b)

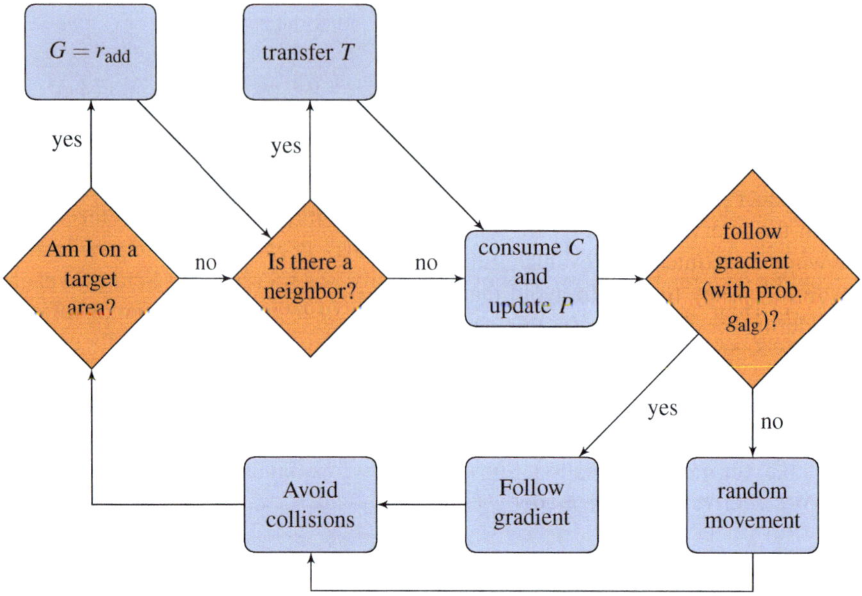

Fig. 5.11 Schematic presentation of the trophallaxis control algorithm. The gathering (G), transfer (T), consumption (C), potential field P, and the ratio of gradient ascent moves (g_{alg}) are defined by Equations 5.14 to 5.18 (Hamann and Wörn, 2008b).

The simulated robots are able to communicate within a short range and with limited orientations, see Section 5.1.1 on page 78 and Schmickl et al. (2007a). By communicating an approximate bearing of the communication partner is determined. It is assumed that infrared devices are used for communication. Additionally, the robots are capable of perceiving certain features of the environment which enables them to determine whether they are on one of the target areas. The explicit feature and the correspondent sensor is not specified in (Schmickl et al., 2007a) and could be, for example, a light sensor or a microphone depending on how the target area is marked.

The fundamental idea of the trophallaxis strategy is to build up a virtual potential field $P(\mathbf{R}_i, t)$ by assigning a value to each robot. This value is stored on each robot i at position $\mathbf{R}_i$ and time t. If the swarm develops a potential field, with maxima at the target areas, robots can, in principle, be lead to the target by a gradient ascent. This approach requires that a connection exists between the specific robot and the robots on the target area (via a low-level multi-hop communication, as described below). In the following paragraph we define an update rule that is applied by each robot to compute $P(\mathbf{R}_i, t)$.

If a robot i is within a target area it adds an amount r_{add} to its P-value (see Figure 5.11 showing the simplified control algorithm). We call G the gathering amount and define

Table 5.2 Model Parameters, distances are given in robot diameters d and time dependent values per simulation step Δt

Parameter	Value
average robot density $\bar{\rho}$	$0.117188/d^2$
number of robots	375
arena dimensions	$80d \times 40d$
target area radii	$1d$ and $5d$
robot diameter	$1d$
communication distance d_{comDist}	$3.5d$
avoiding distance $d_{\text{avoidDist}}$	$0.75d$
robot speed	$0.25d$
addition rate r_{add}	$50/\Delta t$
consumption rate r_{consmpt}	$0.01/\Delta t$
transfer rate r_{transfer}	$0.5/\Delta t$
aggregation threshold δ_{aggr}	0

$$G(\mathbf{R}_i,t) = \begin{cases} r_{\text{add}} & \text{if } \mathbf{R}_i(t) \text{ is within a target area} \\ 0 & \text{else} \end{cases}, \tag{5.14}$$

for addition rate r_{add}. See Table 5.2 for the standard values used in the following examples, if they are not explicitly stated.

To homogenize the potential field a robot i communicates with its neighbors $\mathcal{N}$ (robots within sensor range) and sends or receives a transfer amount T that is proportional to the difference of the potential field values:

$$T(\mathbf{R}_i,t) = \sum_{j\in\mathcal{N}} (P(\mathbf{R}_i,t) - P(\mathbf{R}_j,t))r_{\text{transfer}}, \tag{5.15}$$

for the position of a neighbor $j \in \mathcal{N}$ at $\mathbf{R}_j$ being within communication distance d_{comDist} of robot i and transfer rate r_{transfer}. During each cycle a robot i consumes an amount C that is proportional to P. C is the consumption amount which is defined by

$$C(\mathbf{R}_i,t) = P(\mathbf{R}_i,t)r_{\text{consmpt}}, \tag{5.16}$$

for consumption rate r_{consmpt}.

The update rule for the potential field value of robot i is defined by

$$P(\mathbf{R}_i,t+\Delta t) = P(\mathbf{R}_i,t) + G(\mathbf{R}_i,t) + T(\mathbf{R}_i,t) - C(\mathbf{R}_i,t). \tag{5.17}$$

Before we define the actual rule which determines the robots' movements, we consider the two possible extremes of information being available to a robot in certain situations. In the case of an isolated robot, that is, the robot has no neighbors, and provided the robot has no knowledge about the past and stopping is not an option, all possible directions are equally useful. Thus, turning to a random direction and moving forward for some time to gain a maximum of exploration is a good

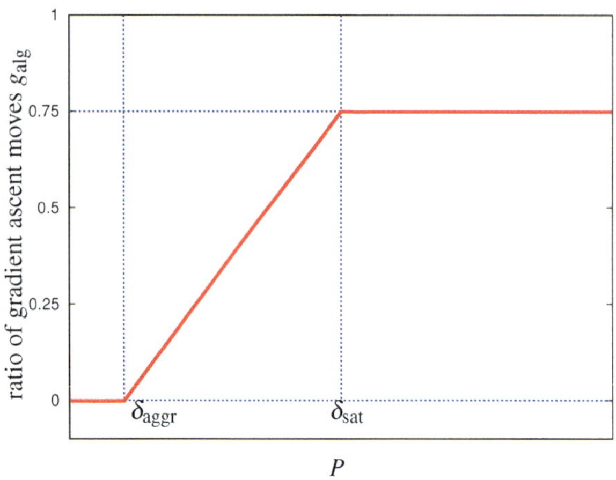

Fig. 5.12 The ratio of gradient ascent moves g_{alg} depending on potential P as defined by the algorithm. For $P < \delta_{aggr}$ all moves are random, for $\delta_{aggr} < P < \delta_{sat}$ the probability of following the gradient increases linearly in P, and for $P > \delta_{sat}$ a robot follows the gradient in 75% of all moves (Hamann and Wörn, 2008b).

strategy for this robot. Concerning the other extreme, a robot having neighbors with high values of P at similar bearings, it is reasonable to move towards them in most of the cases. For all situations between these two extremes, an efficient strategy is less obvious. The movements are described by three phases depending on $P(\mathbf{R}_i,t)$ as shown in Figure 5.12. First, a robot moves randomly if $P(\mathbf{R}_i,t) < \delta_{aggr}$, for an aggregation threshold δ_{aggr}. Second, the ratio g_{alg} of gradient ascent movements to random movements increases linear in P for $\delta_{aggr} < P(\mathbf{R}_i,t) < \delta_{sat}$, for a saturation threshold δ_{sat}. Third, this ratio g_{alg} is constant for $P > \delta_{sat}$. Now, we give a formal definition

$$g_{alg}(\mathbf{R}_i,t) = \max\left(\min\left(\frac{P(\mathbf{R}_i,t) - \delta_{aggr}}{1000}, 0.75\right), 0\right). \tag{5.18}$$

The saturation threshold δ_{sat} is implicitly defined by setting $P(\mathbf{R}_i = \delta_{sat}$ and by applying this to the limiting condition of the min-function argument

$$\frac{\delta_{sat} - \delta_{aggr}}{1000} = 0.75 \Leftrightarrow \delta_{sat} = 750 + \delta_{aggr} \tag{5.19}$$

This approach ensures a minimal exploration rate of 25% independent of P because $g_{alg} \leq 0.75$.

5.2.2 *Model*

Modeling this collective perception scenario with the Langevin and the Fokker–Planck equation is mainly achieved by finding functions $\mathbf{A}$ and B which describe the behavior well. Additionally, we need an equation which describes the temporal evolution of potential P. Defining these functions is a creative process and there is no unique solution.

Function $\mathbf{A}$ depends primarily on a vector representing the directional information, that is, the gradient ∇P. We normalize this gradient because the robots are not influenced by the vector length of ∇P but by the direction only. The frequency of choosing the direction given by ∇P depends on the gradient ascent ratio g_{alg}. We use an effective gradient ascent ratio g_{mod} differing from g_{alg} due to macroscopic effects as described later. Furthermore, we have to define a maximal density, otherwise, the whole swarm could aggregate in one point. This can be incorporated in a discount coefficient $d(\rho) \in [0,1]$ with $d(0) = 1$ and $d(\rho_{\max}) = 0$ for a maximal density $\rho_{\max}$. By d the increasing difficulty of robots executing gradient ascent movements with increasing density is modeled. High densities result in less successful executions of gradient ascent moves because more collision avoidance actions are necessary. Finally, we multiply with the nominal velocity v because, in terms of the Fokker–Planck equation, the robot flow depends linearly on the velocity. In terms of the Langevin equation: The robot's displacement in direction ∇P is in the average a fraction of v. If all gradient ascent moves would be successful ($d = 1$) and $g_{\text{mod}} = 1$ the displacement would be exactly v. However, from time to time a gradient ascent move is prevented due to interfering neighbors ($d \neq 1$) or an intended exploration move ($g_{\text{mod}} < 1$) is done. Thus, we get

$$\mathbf{A} = d(\rho) g_{\text{mod}}(P) \frac{\nabla P}{\|\nabla P\|} v, \tag{5.20}$$

and we set $\mathbf{A} = 0$, if $\|\nabla P\| = 0$. Function B describes the nondeterministic motion. In correspondence to the definition of $\mathbf{A}$, random movements are either caused by intended exploration with a rate of $(1 - g_{\text{mod}})$ or by a gradient ascent attempt disturbed by collision avoidance with a rate of $(1-d)g_{\text{mod}}$. Additionally, we have to model a slowdown of the robot's random movements. We want to underline: Due to the same cause (increasing density) we observe two consequences in the swarm behavior. First, the number of collisions increases. Second, the average displacement due to collision avoidance actions decreases because the collision avoidance is itself interrupted by approaching other robots. We can reuse d as a good description of the normalized spatial displacement since the displacement is inverse proportional to the density (analogous to the mean free path in physics). Thus, multiplying by d and velocity v we get

$$B = \big((1 - g_{\text{mod}}) + (1 - d) g_{\text{mod}}\big) v d = (1 - g_{\text{mod}} d) v d. \tag{5.21}$$

In the following, we model the communication of the robots, that is, the dynamics of the potential field. Unfortunately, there is no exactly defined and physically motivated way of deriving a partial differential equation for the communication as it exists for the robots' motion. However, the characteristics of the scenario addressed here allow an intuitive derivation.

At first, consider the transferring process. A robot determines the difference between its own potential field value and the value of a neighbor that is within communication range. Then a part of this difference defined by r_{transfer} is transfered to the robot with the smaller value. This exactly corresponds to the implementation of the Laplace operator ($\Delta = \nabla^2$, used to model diffusion processes) using the finite difference method with a coefficient. However, this coefficient is hard to determine. It can be seen as a diffusion constant but in the simulation the distance between transferring robots has no effect to the amount that is transferred as long as they are in communication range. We can compute an effective coefficient c_{eff} for a distance d between the two communicating robots, which is given by $c_{\text{eff}} = c^{1/d}$. By this definition we get $c_{\text{eff}}^d = c$, which corresponds to d multiplications of c_{eff} we would have in grid discretization with a grid point distance of one unit. Additionally, robots can transfer their amount of the potential field just by moving around which is also a diffusion process describable by the Laplace operator. The selected definition of the transfer coefficient is motivated by the above reasoning and was empirically approved:

$$c_{\text{transfer}} = r_{\text{transfer}}^{1/d_{\text{comDist}}}. \tag{5.22}$$

By this definition, we ensure that the effective coefficient for distance d_{comDist} is the transfer rate c_{transfer}.

Modeling the consumption process is straight forward. Since a single robot consumes $r_{\text{consmpt}} P(\mathbf{R},t)$, the consumption at a position $\mathbf{r}$ is defined by $r_{\text{consmpt}} P(\mathbf{r},t)\rho(\mathbf{r},t)$ because ρ gives the expected number of robots as defined above. With the same reasoning the addition process is modeled. Hence, the PDE for the potential field is given by

$$\begin{aligned} \frac{\partial}{\partial t} P(\mathbf{r},t) = \quad & c_{\text{transfer}} \nabla^2 P(\mathbf{r},t) \\ & - r_{\text{consmpt}} P(\mathbf{r},t)\rho(\mathbf{r},t) \\ & + \delta_{\text{onTarget}}(\mathbf{r},t) r_{\text{add}} \rho(\mathbf{r},t), \end{aligned} \tag{5.23}$$

with

$$\delta_{\text{onTarget}}(\mathbf{r},t) = \begin{cases} 1 & \text{if } \mathbf{r} \text{ is within a target area at time } t \\ 0 & \text{else} \end{cases}. \tag{5.24}$$

The prediction of this model is only qualitatively correct because we assume that the limited amount added to the system is dispersed over the whole area. But in the microscopic simulation it is distributed in singularities leading to higher values than in our model. However, it turns out that at least for sufficiently high densities (about $\rho > 0.07$) there is a constant relationship between the potential field in the model and in the simulation. We get a quantitative model by multiplying the potential field values of the model with an empirically determined scaling factor. This factor is determined by drawing samples of the potential field from the simulation and dividing them by the prediction of the un-scaled model. The need of such a free parameter is, in general, dissatisfying. However, here it needs to be measured just once for the scenario. There is always a tradeoff between the complexity of the model and the need for empirical parameters. On the other hand the accuracy of the model, as presented in the next section, without an intensive optimization of the empirical parameters shows the insensitivity of the model to these values. We explain next, how we determined the parameters of this model.

Measured and Free Parameters

The discount coefficient function $d(\rho)$ can be measured by counting the number of collision-avoiding actions of a robot group in an otherwise empty arena. This function's principle shape is easily diagnosed. However, it is difficult to determine the function quantitatively. In the control algorithm, virtual physics is used to avoid collisions, that is, the deflection is highly variable, and therefore, the difference between collision avoidance and regular moves is blurred. We decide to consider a deflection of at least $\pm 20^\circ$ as an avoidance move. d is the ratio of such avoidance moves to the total number of moves. We obtained $d(\rho) = (1-\rho)^{3.7}$ as a good fit for the measurements, see Figure 5.13. Consequently, we have reduced the complex and microscopic underlying processes to a single parameter which does express few. However, this abstraction of microscopic processes is necessary and we readily accept it because it is measurable.

The effective gradient ascent ratio g_{mod} is also based on measurements. We measured mean potential field values and corresponding mean values g_{eff} for 13 histogram bins on the interval $g_{\text{eff}} \in [0, 1200]$ over 600 simulation runs of 250 time steps each, see Figure 5.14 on the next page. In doing so, we use internal information, something that is impossible in biological systems, that is, it cannot be determined based on the behavior. We get a good measure for the frequency of goal-oriented actions as a function of P. Using g_{mod}, which is an approximation for

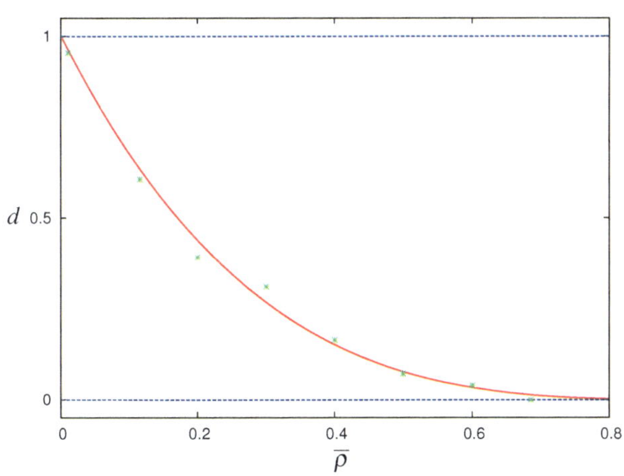

Fig. 5.13 Discount coefficient d, asterisks show measured values; confidence intervals are within the symbols' sizes (Hamann and Wörn, 2008b).

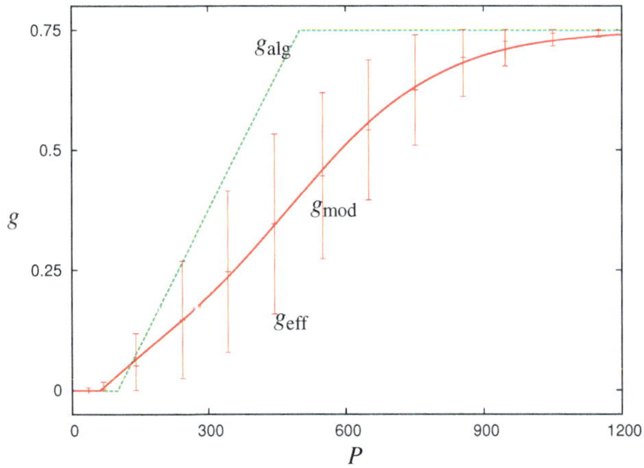

Fig. 5.14 Ratio of gradient ascent moves g_{mod} (solid line) for $\delta_{aggr} = 100$ depending on potential P as used in the model in comparison to measured values g_{eff} (error bars show the standard deviation) and to g_{alg} (dashed line) as defined in the algorithm (Hamann and Wörn, 2008b).

g_{eff}, we include the high variance of P in the model. Thereby, the simplification of total connectivity is compensated as well because this variance is mainly caused by alternating between connection and disconnection of robot groups. In addition, it turns out that the influence of the aggregation threshold δ_{aggr} creates only a shift allowing for the generalization from g_{eff} to any δ_{aggr}.

Although several parameters can be measured, some free parameters remain due to our empiric approach of the communication model. While the gathered potential field value in the simulation is just distributed among the robots, it is dispersed over the whole area in our model. Therefore, we scale the height of the potential field by a free coefficient. Also, the absolute displacement caused by a collision (Q in Equation 4.66 on page 64) cannot be measured because one avoidance event cannot be distinguished from the next one. The scaling parameter and Q were determined by fitting the model to the simulation data of Schmickl et al. (2007a) shown in Figure 5.15 on the next page. The resulting single parameter setting was used to obtain all other results, which will be described in the following section.

5.2.3 Results and Discussion

We verify the model by focusing on three parameters (target radii, average robot density $\overline{\rho}$, and aggregation threshold δ_{aggr}) and by comparing the predictions of our model to the data obtained in simulations reported in (Schmickl et al., 2007a). While the free parameters were fitted to the data of Figure 5.15 on the facing page, the other data can definitely be considered as predictions of the model: number of aggregated robots at the target areas depending on the robot density

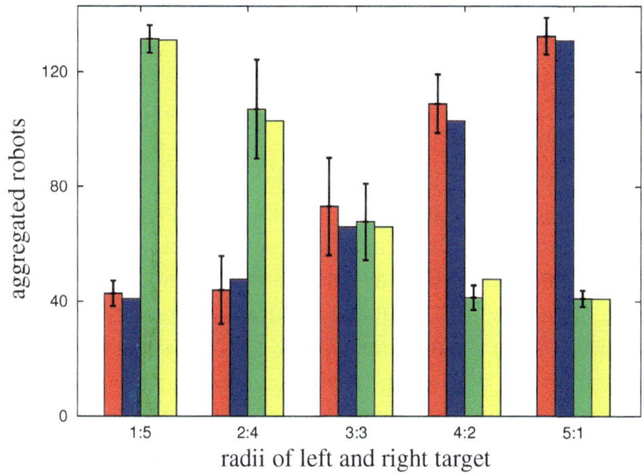

Fig. 5.15 Number of aggregated robots with varied target area radii, comparing the simulation of Schmickl et al. (2007a) to the model: aggregated robots at left target in simulation (1st bar of each cluster) and model (2nd bar), aggregated robots at right target in simulation (3rd bar) and model (4th bar), $\overline{\rho} = 0.13$ (Hamann and Wörn, 2008b).

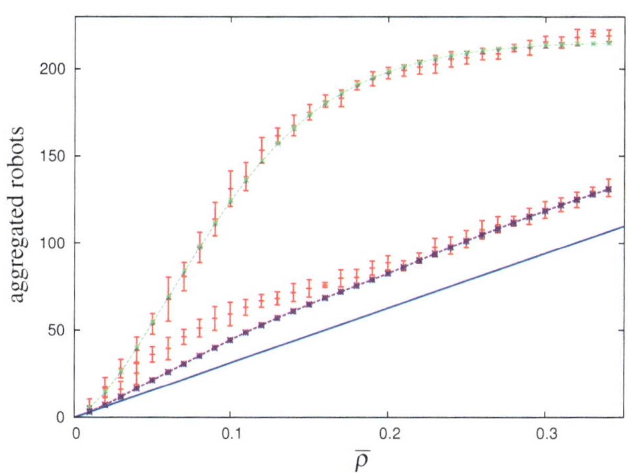

Fig. 5.16 Number of aggregated robots at the small (lower line) and the big target area (upper line) with varied average density $\overline{\rho}$, comparing the model (dashed lines) to the simulation (confidence intervals). The solid line gives the expected number of robots if there is no aggregation, $\delta_{\text{aggr}} = -50$ (Hamann and Wörn, 2008b).

(Figure 5.16), number of aggregated robots at the target areas depending on the aggregation threshold δ_{aggr} (Figure 5.17 on the following page), and the robot density over space (Figure 5.18 on page 99). Therefore, we have a good fit in Figure 5.15 while we have more significant discrepancies in the other figures. The accuracy of these predictions is sufficient such that the model is able to support the algorithm design phase. For example, an appropriate aggregation threshold δ_{aggr} can be chosen on the basis of Figure 5.17 which shows a balance between occupying the bigger target area and regarding the smaller area somewhat.

The systematic deviations in Figure 5.16 on the preceding page and 5.17 (underestimation of aggregated robots at the smaller target area for $0.03 < \overline{\rho} < 0.2$ and $-200 < \delta_{aggr} < 0$) seem to be caused by temporal breakdowns of the coherence around the smaller target area. Such breakdowns are more likely for smaller robot densities. Our observations of the simulations show that small, isolated groups of robots are able to establish a potential field with a significant gradient at the smaller target area for a limited time. This gradient suffices to cause a temporary aggregation. Our approach to compensate such fluctuations with g_{mod} is not sufficient in the case of smaller robot densities because we averaged over the time by measuring g_{eff}. Thus, for simplicity reasons, we assumed statistical independence over time. However, our observations show a correlation with time. Thus, higher accuracy in the collective perception scenario seems to be impossible based on the current model state (positions but no directions) and the underlying Markov approximation. An even more sophisticated approach is apparently needed to overcome this problem.

In Figure 5.18 on the facing page we compare the spatial robot densities of the simulation and the model. The squared error map in Figure 5.18(c) on the next page indicates an incorrect prediction of the density shape at the bigger target area. This shape is very sensitive to the discount coefficient d (Figure 5.13 on page 95). However, the mean squared error is quite low (≈ 0.001 for a maximal value in the densities of 0.035).

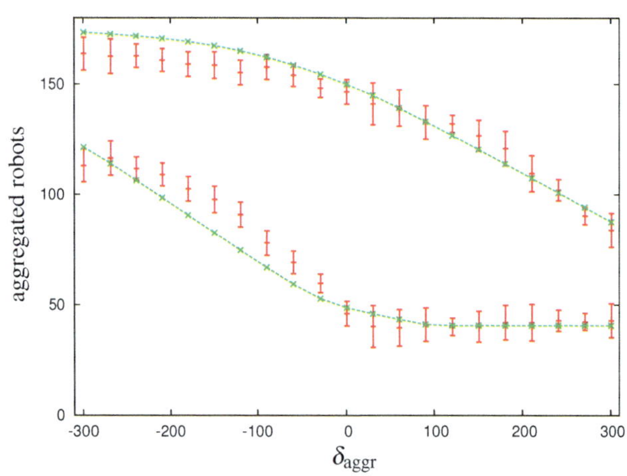

Fig. 5.17 Number of aggregated robots at the small (lower line) and the big target area (upper line) depending on aggregation threshold δ_{aggr}, comparing the model (dashed lines) to the simulation (confidence intervals, Hamann and Wörn, 2008b)

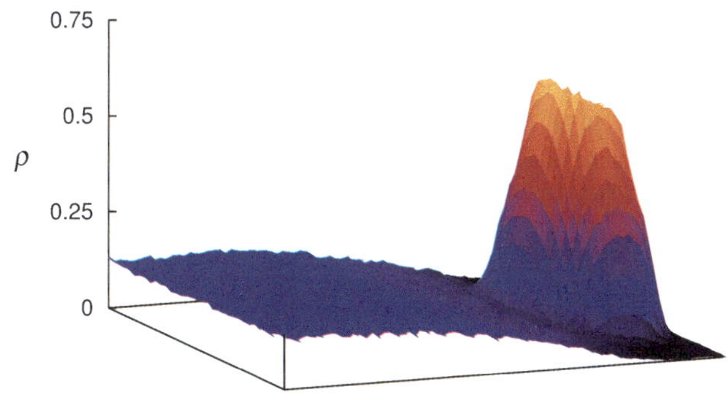

(a) Simulation, mean of 600 samples.

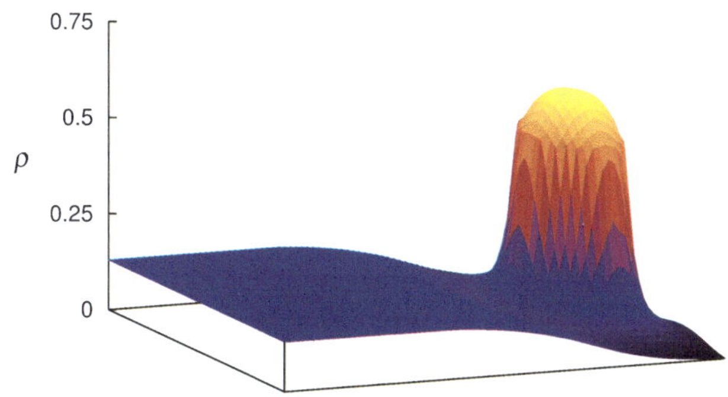

(b) Model.

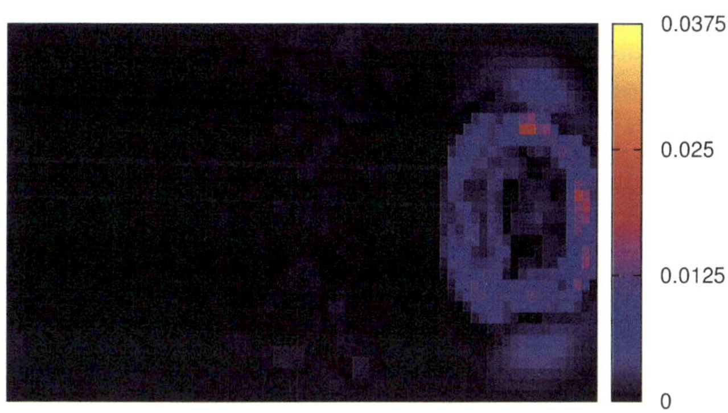

(c) Squared error, maximum about 0.035.

Fig. 5.18 Comparison of the robot density $\rho(\mathbf{r}, t = 250)$ between simulation and model (Hamann and Wörn, 2008b)

5.3 Emergent Taxis

The case study reported in this section was previously published in Hamann and Wörn (2008b). We give a qualitative model of the emergent taxis scenario as presented by Nembrini et al. (2002) and analyzed by Bjerknes et al. (2007); Nembrini and Winfield (2010). This algorithm is a great example of how a comparatively complex task, moving collectively towards a beacon, can be solved by a simplistic approach. In the algorithm there is no instruction telling a robot to move towards the beacon, which is possibly why it is labeled "emergent".

5.3.1 *Scenario*

The primary task of the robot swarm is to move towards a beacon. Secondary tasks are the avoidance of collisions and to preserve coherence, that is, to keep a connection with a predefined number of neighbors. All robots have a range-limited wireless communication device. Each robot frequently sends a ping and permanently counts the individual robots from which it receives such a ping. All robots move forward until one of the following three events occurs (cf. Figure 5.19):

1. If two robots get too close to each other they have to avoid the collision.
2. Robots receiving pings from too few neighbors try to regain coherence.
3. Robots having recently regained coherence change their direction randomly.

The following process ensures coherence: If the number of individual robots, from which a ping was recently received, drops below a threshold α (condition for the transition to the *coherence* state), the robot executes a U-turn and re-enters the *forward* state. The underlying assumption is that the robot was about to detach and lose connection to the swarm. The U-turn followed by moving forward will reconnect the robot. After the number has reached the threshold α again (condition for the transition to the *random* state), the robot makes a random turn and re-enters the

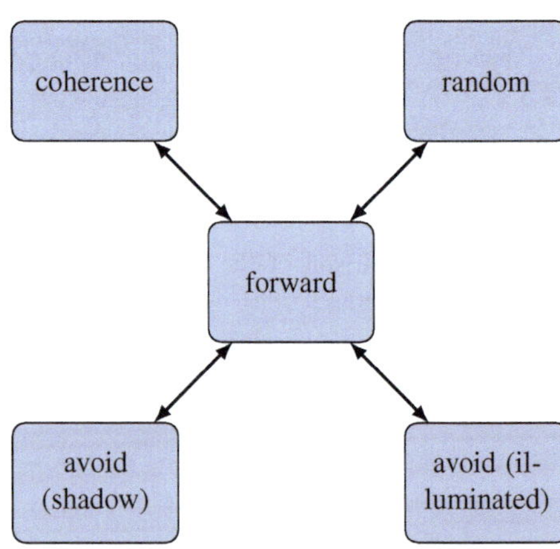

Fig. 5.19 State diagram for the emergent taxis scenario, reproduced after (Bjerknes et al., 2007)

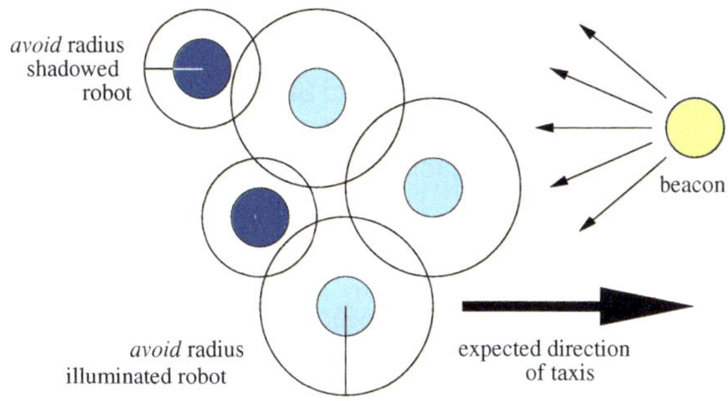

Fig. 5.20 The emergent taxis scenario with five robots (dark blue circles for shadowed and light blue for illuminated robots) and a beacon to the right (yellow circle). Circles around the robots show the avoid radii. The communication range is not shown (Hamann and Wörn, 2008b).

forward state. In the simulations given by Bjerknes et al. (2007) α is always set to the swarm size $N = \alpha$. The two *avoid* states implement the collision avoidance behavior. In both states, the robots turn away from other robots which were getting too close. The difference of these states is discussed in the following.

Up to this point the algorithm triggers an aggregation behavior (ideally) ensuring the swarm stays connected. To extend this behavior by a collective movement towards the beacon, each robot is equipped with a long-range beacon sensor enabling it to sense whether there is a beacon. It is assumed that occlusions are possible, that is, a robot which is in the shadow of another robot will not perceive the beacon (cf. Figure 5.20). In the following, we distinguish two groups of robots: illuminated (perceiving the beacon) and shadowed (not perceiving the beacon). The crucial point of the emergent taxis algorithm is to define different *avoid* radii for these two groups. Illuminated robots have a bigger *avoid* radius than shadowed robots. This is the difference between the two *avoid* states which is enough to determine the desired direction of the taxis.

5.3.2 Model

Now, we want to model the fundamental behavior. Two insights about the macroscopic process are helpful to obtain a simple model. First, we note the symmetry breaking: illuminated robots avoid shadowed robots but the reverse does not occur — as if shadowed robots would not perceive illuminated ones. Thus, the goal-oriented movements of illuminated robots (behavior implemented by the *coherence* state) will be interrupted more often than the movements of shadowed robots. This effect will result in a higher ratio of random motion for illuminated robots compared to shadowed robots. Furthermore, the asymmetric *avoid* radii result in a decreased number of collisions for shadowed robots. Stated differently, the effective robot density for shadowed robots will be lower then the actual robot density as it is perceived by and effective to the illuminated robots. While illuminated robots have to avoid

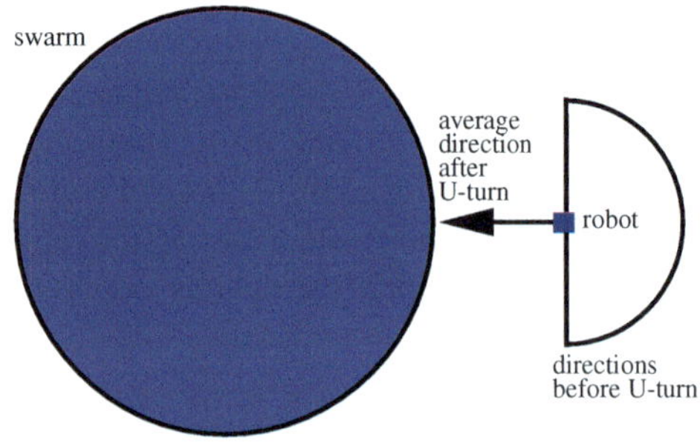

Fig. 5.21 Schematic representation of a U-turning robot that was about to detach from the swarm, indicating the average direction towards the swarm's barycenter after the U-turn (Hamann and Wörn, 2008b)

shadowed robots, shadowed robots will not avoid illuminated robots because the illuminated robots have to react earlier, that is, at bigger distances. Second, we assume that after having executed their U-turn, robots will, on average, head towards the barycenter of the swarm (see Figure 5.21 showing a detaching robot).

Given these insights, the adaptation of the model framework to the emergent taxis scenario is straightforward. At first, we have to introduce the distinction between illuminated and shadowed robots. Considering the probabilistic nature of our model we introduce function $s(\mathbf{r},t)$, giving the probability of a robot at position $\mathbf{r}$ and time t of being shadowed. The values of s are achieved by integrating over ρ along beams originating in the beacon (ρ is a particle density, that is, integrating over all space we get the swarm size $N = \int \rho(\mathbf{r})d\mathbf{r}$). To ensure the probability property, we define a maximum of $s_{\max} = 1$. We define for the general case

$$s(\mathbf{r},t) = \min\left(\int\limits_{r'_y = mr'_x + k} \rho(\mathbf{r}',t)d\mathbf{r}', 1\right), \tag{5.25}$$

with $\mathbf{r}' = (r'_x, r'_y)^T$ and for a light source at $\mathbf{L} = (L_x, L_y)^T$. We define $m = \frac{r_y - L_y}{r_x - L_x}$ and $k = r_y - \frac{r_y - L_y}{r_x - L_x} r_x$ for $\mathbf{r} = (r_x, r_y)^T$. Here, we assume that the light source is far away in the x direction. Thus, a simplified light model can be applied assuming the light beams arrive parallel to the x-axis (cf. Figure 5.22 on the facing page). This simplifies Equation 5.25 to

$$s(\mathbf{r},t) = \min\left(\int_{r_x}^{\infty} \rho(\mathbf{r},t)d\mathbf{r}', 1\right), \tag{5.26}$$

for $L_x \rightarrow \infty$. An intuitive interpretation of this equation is to think of summing densities beginning at the light source backwards in the x direction. If a point is reached, where the value sums up to one, we set $s = 1$. Otherwise, s is set to the sum reached thus far.

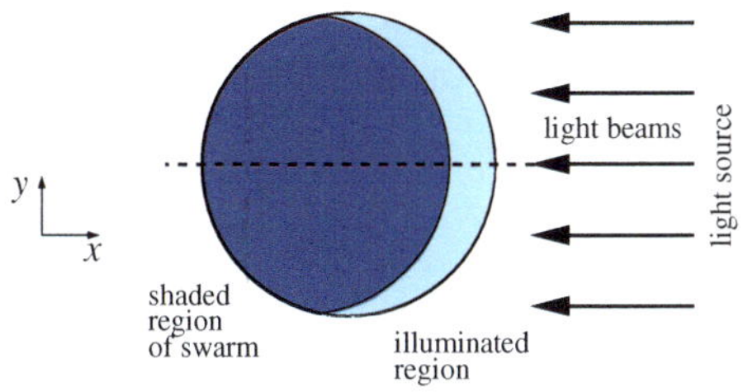

Fig. 5.22 Schematic representation of the light model used. The light beams are assumed to be parallel. On the right hand side of the swarm a region with $s > 0$ is found (illuminated region). On the left hand side there is a region with $s = 0$ (shadowed region). The actual size of the illuminated region depends on the swarm size, robot size/shape etc. Integration over ρ along lines parallel to the dashed line determines s, cf. Equation 5.25 on the preceding page (Hamann and Wörn, 2008b).

Based on s we define the diffusion function B as a sigmoid function

$$B(\mathbf{r}(t),t) = (1 + \exp(-c_1\rho(\mathbf{r},t) + (1 - s(\mathbf{r},t))c_2 + s(\mathbf{r},t)c_3))^{-1}, \tag{5.27}$$

for constants c_1, c_2, and c_3 (see Table 5.3 for the values that were used in the following example).

Table 5.3 Parameters used in the taxis scenario, by u the unit length is indicated

Parameter	Value
N	10
c_1	$50[1/u^2]$
c_2	7
c_3	11
c_4	$50[1/u^2]$
c_5	23 (for a_1) and 30 (for a_2)

The gradient function $\mathbf{A}$ represents the threshold α and the direction of robots after having executed a U-turn. This representation is achieved by defining a potential $P(\mathbf{r},t)$ which has a global maximum at the barycenter of the swarm at each time and whose gradient is normalized $\forall \mathbf{r},t : \|\nabla P(\mathbf{r},t)\| = 1$. Combining this gradient with a sigmoid function which activates the U-turning behavior for low densities, we get

$$\mathbf{A}(\mathbf{r}(t),t) = a(\mathbf{r}(t),t)\nabla P(\mathbf{r},t), \tag{5.28}$$

with

$$a(\mathbf{r}(t),t) = (1 - (1 + \exp(-c_4\rho(\mathbf{r}(t),t) + c_5))^{-1}), \tag{5.29}$$

for constants c_4 and c_5. The parameters that were used in the following examples are given in Table 5.3 and plots of B and a are given in Figure 5.23 on the following page.

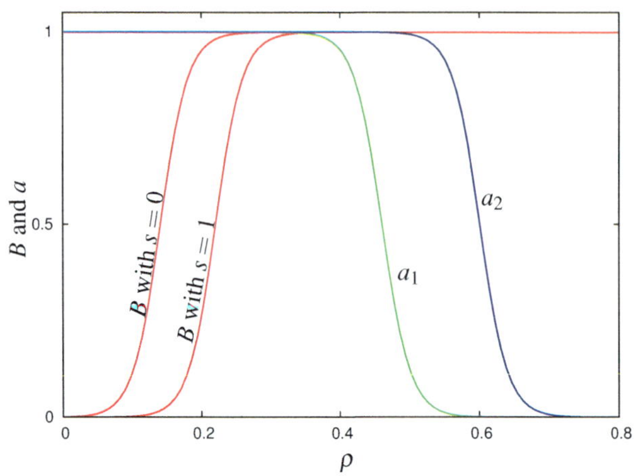

Fig. 5.23 Examples for sigmoid functions a and B for $s = 0$ and $s = 1$ (Hamann and Wörn, 2008b)

5.3.3 *Results*

The obtained simple model is already capable of predicting the velocity of the swarm's collective motion in a qualitatively correctly manner (change of the barycenter) as is shown in Figure 5.24 which gives the distance covered by the swarm. The difference between the two runs is a shifted a-function (a_1 and a_2). Higher swarm velocities are achieved by shifting a to lower densities, that is, robots can reach bigger distances from the barycenter (where the robot density gets

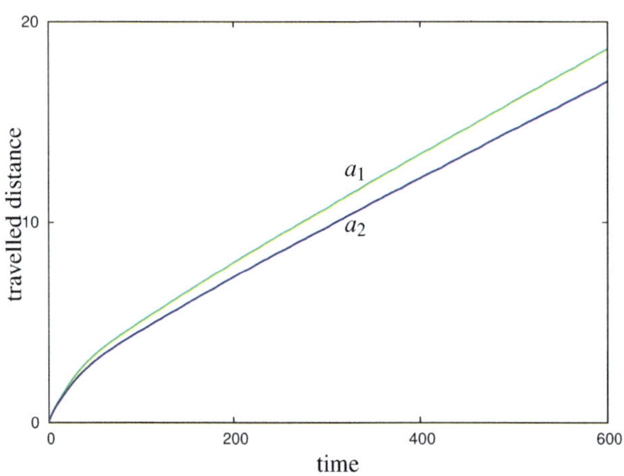

Fig. 5.24 Distance covered by the swarm collectively measured on the basis of the barycenter for two variants of function a (cf. Figure 5.23, Hamann and Wörn, 2008b)

lower) before they have to do a U-turn. This effect corresponds to the observations by Bjerknes et al. (2007). They observed higher swarm velocities for bigger communication distances. Thus, our model's prediction is a qualitative result in accordance with Bjerknes et al. (2007).

5.4 Foraging

The case study reported in this section was previously published in Hamann and Wörn (2007a). The swarm task is to explore the arena in search for an area called *food* place. We assume, the robots pick up objects there that need to be transported to an area called *nest*. The robots change their local environment while moving. This is done in order to inform other robots, that are passing by later, about the positions of food and nest. Here, this is done by virtual pheromones. A fully convincing implementation of virtual pheromones is still pending. However, there are two main options such as: projections of images onto the arena that are perceived and indirectly manipulated by the robots (making a central device necessary, Sugawara et al. (2004)) and implicit directional information that is communicated (Payton et al., 2001; Schmickl et al., 2007a). In the following, the actual implementation of virtual pheromones is not addressed. Instead, the focus is on the modeling two states with two robot densities.

5.4.1 Scenario

As a framework for our simulation we use the Breve simulator by Klein (2003). Here, we simulate a homogeneous swarm in continuous space combined with a discrete implementation of the pheromones. Since we want to simulate big numbers of robots over many runs, we depend on a rather simple model of the robot, that is computationally easy to handle. In our model, the robot has circumferential visibility, can measure distances to objects within the coverage of its sensors, and can distinguish between other robots, the nest, food, and the wall, that circumscribes the arena. It is also able to perceive a pheromone gradient in two mutually orthogonal directions and to drop a certain amount of pheromone. The robots' locomotion is assumed to be ideal, that is, an acceleration towards an arbitrary direction is possible at all times (holonomic drive mechanism). The control of the robot is totally reactive and based on the principle of virtual physics (potential field techniques), that is, other objects have a repelling effect on it depending on their distance and visibility (Spears and Gordon, 1999; Khatib, 1986; Arkin, 1989). This defines the avoidance behavior to be similar to the collision of two particles in our real physical world.

A robot is in one of two possible states: looking for food (s_f) or returning home to the nest (s_n). Initially, all robots are randomly positioned close to the nest with a random velocity heading to a random direction, and starting in state s_f. If a robot in state s_f perceives the food, it switches to state s_n. Robots in state s_n perform a transition to state s_f, if they perceive the nest. We are using two pheromones: one

Table 5.4 Simulation parameters

Parameter	Value
arena size	258 cm × 258 cm
nest position	(129 cm, 195 cm)
food position	(129 cm, 63 cm)
patch size	6 cm
agent diameter	2 cm
proximity sensor range	5 cm
iteration step	0.05 s
max. speed	7 cm/s
evaporation rate	0.0392 1/s
diffusion rate	0.1568 1/s
drop decrease rate	0.095 1/s
initial drop size	0.5

that should be established to increase in intensity towards the food (p_f) and another one that increases towards the nest (p_n). To avoid immense instabilities and to simplify the scenario, pheromone p_n is chosen to be present and constant at all times. It is always guaranteed to have a smooth gradient leading to the nest at any position. However, at least in a grid world based on a concept of dying agents it has been shown, that a stable behavior can be reached with two dynamic pheromones with the advantage of finding shortest paths around obstacles (Panait and Luke, 2004). As a second consequence, the robots will only be able to deposit pheromone p_f, which is done in state s_n. The amount that is dropped by the robot at each simulation step is set to an initial value (drop size, see Table 5.4) at the state transition and decreases exponentially over time thereafter (drop decrease rate).

The pheromones are implemented by a grid of so-called patches covering the whole arena. The patches are quadratic with side length $s = 6cm$ (for comparison: a patch fits into the area covered by the robot's sensors). The performance of the swarm is independent of this size as long as it is reasonably small and both the evaporation and the diffusion rate are adapted to it. But choosing the patch size is computationally critical because the evaporation and diffusion process of pheromone p_f is executed at every time step, which has to handle every single patch (complexity is $O(\frac{1}{s^2})$), another option could be to update the grid only every m time steps for $m > 1$). Every patch has an associated pheromone intensity i that is updated per step by

$$i_{t+1} = (1-e-d)i_t + \sum_{n \in \mathcal{N}} n(1-e)\frac{d}{4}, \tag{5.30}$$

where e denotes the evaporation rate, d the diffusion rate and $\mathcal{N}$ the set of intensities of the patches in the von Neumann neighborhood of range one, except the current patch itself. If a robot deposits some pheromone p_f, the dropped amount will be added to the intensity associated with the patch where the robot is located at that moment. Pheromone p_n is time-invariant as discussed above (see Section 5.4.3 on page 109 for the definition).

To implement the gradient ascend, the two components of the gradient are computed from the intensities of the neighboring patches independent of the robot's orientation: $g_x = n_{x+} - n_{x-}$, where n_{x+} denotes the intensity of the neighboring patch in positive x-direction and n_{x-} in negative x-direction; g_y is computed analogously. The overall acceleration vector of the robot is a weighted sum of $\frac{(g_x,g_y)^T}{\|(g_x,g_y)^T\|}$ (if $\|(g_x,g_y)^T\| = 0$ the term is set to 0) and another vector depending on sensed objects implementing the avoidance behavior.

5.4.2 Model

We define two probability densities: One for robots in state s_f and one for robots in state s_n. Thus, we apply Equation 4.71 on page 67 again and get

$$\frac{\partial \rho_f}{\partial t} = -\nabla(\mathbf{A}_f \rho_f) + \frac{1}{2}\nabla^2(B_f^2 \rho_f) + t_{n,f}\rho_n - t_{f,n}\rho_f, \tag{5.31}$$

$$\frac{\partial \rho_n}{\partial t} = -\nabla(\mathbf{A}_n \rho_n) + \frac{1}{2}\nabla^2(B_n^2 \rho_n) + t_{f,n}\rho_f - t_{n,f}\rho_n, \tag{5.32}$$

with

$$\mathbf{A}_f(\mathbf{r},t) = \alpha_f \frac{\nabla P_f(\mathbf{r},t)}{\|\nabla P_f(\mathbf{r},t)\|}, \tag{5.33}$$

and $\mathbf{A}_n$ accordingly. The state transitions $t_{f,n}$ and $t_{n,f}$ are space-dependent and mostly set to $t_{f,n} = t_{n,f} = 0$ except for the nest and the food area. $t_{n,f}$ is non-zero at the nest and $t_{f,n}$ is non-zero at the food. The state transitions are modeled as special boundary conditions. In the following, we give the definition for $t_{n,f}$, while $t_{f,n}$ is not discussed here but defined accordingly. Say $\partial\Omega_n$ is the boundary of the arena around the nest. Then we define the boundary conditions at the nest as the following:

$$\begin{aligned}\forall \mathbf{r} \in \partial\Omega_n : \frac{\partial}{\partial t}\rho_f(\mathbf{r},t) = &- \nabla(\mathbf{A}_f(\mathbf{r},t)\rho_f(\mathbf{r},t)) + \frac{1}{2}\nabla^2(B_f^2(\mathbf{r},t)\rho_f(\mathbf{r},t)) \\ &- \nabla(\mathbf{A}_n(\mathbf{r}-\varepsilon\mathbf{n},t)\rho_n(\mathbf{r}-\varepsilon\mathbf{n},t)) \\ &+ \frac{1}{2}\nabla^2(B_n^2(\mathbf{r}-\varepsilon\mathbf{n},t)\rho_n(\mathbf{r}-\varepsilon\mathbf{n},t))\end{aligned} \tag{5.34}$$

$$\forall \mathbf{r} \in \partial\Omega_n : \rho_n(\mathbf{r},t) = 0, \tag{5.35}$$

where $\mathbf{n}$ denotes the exterior normal to the boundary (pointing towards the nest center). The intuitive interpretation of these equations is simple: The robots in state s_n, that are close to the nest, perform a transition to s_f, because they have finished their mission to find the nest. In a trivial grid discretization of these PDE, the boundary conditions are implemented by adding the amount of ρ_n to ρ_f and setting $\rho_n = 0$

Table 5.5 Pheromone parameters

Parameter	Value
c_1	$1.4 \cdot 10^{-4}$
c_2	0.998
c_3	10

within the area of the nest after each iteration. The boundary conditions at the food are defined in an analog way. The boundaries of the arena are modeled as total isolation.

The pheromone P_n leading to the nest is assumed to be constant over time and is just defined as it is implemented in the simulation:

$$P_n(\mathbf{r}) = c_1(\sqrt{d_{max}} - \sqrt{\text{dist}(\mathbf{r}, \mathbf{r}_n)}), \tag{5.36}$$

where $\text{dist}(\mathbf{r}, \mathbf{r}_n)$ is the distance from $\mathbf{r}$ to the center of the nest $\mathbf{r}_n$, d_{max} is the maximal possible distance, and constant c_1 is used to adapt P_n to the absolute intensities of P_f (see Table 5.5). However, every function that provides a gradient pointing towards the nest at all positions could be used, since our model as well as the simulation are both independent of absolute values.

Pheromone P_f is modeled to depend on ρ_n directly:

$$P_f(\mathbf{r}, t) = \rho_n(\mathbf{r}, t) c_2^{c_3 \text{dist}(\mathbf{r}, \mathbf{r}_f)}, \tag{5.37}$$

where $\text{dist}(\mathbf{r}, \mathbf{r}_f)$ is the distance from $\mathbf{r}$ to the center of the food $\mathbf{r}_f$, and constants $c_2 < 1$ and c_3 depending on the pheromone dropping procedure. The underlying consideration is that the amount of pheromone dropped per step by a robot decreases exponentially with time. Thus, the amount of pheromone that can be dropped by a robot at a certain spot is limited by the time a robot needs to travel from the food to this spot. Assuming a constant velocity this time is proportional to the distance to the food.

Note that no dependency on the history of ρ_n is incorporated. Intuitively, one might argue, that this direct coupling corresponds to a high evaporation rate of the pheromone. Thus, this direct coupling could cause instabilities in the modeled overall behavior of the swarm, that would not emerge for appropriate parameter settings. However, for reasonable values of diffusion B, the history is intrinsically modeled. For the steady state this is quantitatively true and can be shown under the assumption that at each patch the pheromone diffusion net flux is zero. Since the robots depend on the normalized gradient only, a qualitatively correct representation of the gradient is sufficient. For visualization, imagine a large group of robots in state s_n starting at the food and moving towards the nest following the gradient greedily and leaving behind only few other robots that moved to different directions due to diffusion. On the line between food and nest they would create a pheromone trail starting high at the food and decreasing exponentially towards the nest because of the dropping method. Since the density of the robots left behind would not exponentially increase towards the nest, this situation is represented qualitatively correct in our model. Similar considerations imply that Equation 5.37 on the facing page is a good approximation.

5.4.3 Results

To simplify the following investigations, we restrict ourselves to a special case and set $\alpha_f = \alpha_n = \alpha$ and $B_f = B_n = \sqrt{2(1-\alpha)} = B$. This class of parameter settings corresponds to the intuition that an increased amount of goal-oriented motion causes a decrease in the diffusion and vice versa. Now, only a single variable that connects the model to the simulation is left — the diffusion B.

The diffusion has to be determined from the overall behavior in the simulation. This was done by measuring the amount of robots that follow the pheromone gradient approximately in comparison to those that move to any other direction at many different positions in the arena. The average diffusion was computed by weighting these values by their local density averaged over time.

The solution of the PDE provides us basically with the stationary density distributions of robots in states s_f and s_n, if they exist. Figure 5.25(a) on the next page shows a typical solution and Figure 5.25(b) the corresponding averaged density of 30 simulation runs. The accuracy of the model is good at positions between nest and food. However, close to the nest and the food it suffers from the unmodeled acceleration processes of the robots and the impossibility to represent the infinite slope of the densities in the simulation.

Furthermore, we are interested in the resulting flow of food that is the same as the rate of robots that perform a transition from s_n to s_f per time. The amount of robots that perform a transition at the boundaries of the nest and the food per time in our model gives a good prediction of the flow. Another estimate of the converged flow, that turned out to be less sensitive to correctly measuring the diffusion, is obtained by integrating the densities of the steady-state over a line in the middle between nest and food, that is orthogonal to the shortest path between nest and food. Say the result of this integration is I then the prediction of the flow for the given diffusion constant B would be $I(1-(0.5B^2)) = I\alpha$. This is a good approximation because the component of the pheromone gradient in the direction orthogonal to this plane is typically small. Thus, the diffusion in this direction will also be small following Fick's first law that gives the diffusion flux in the steady state: $J(\mathbf{r}) = -B\frac{\partial P(\mathbf{r})}{\partial \mathbf{r}}$. This is the method that was used to predict the flow here.

In every simulation run, we drew 36 samples equidistant in time after a transient. The results are averaged over the samples of 30 runs. More runs would be desirable for a better statistical significance, however, due to limited resources and the high computational demand of the simulation, that could not be achieved. Two sets of simulation runs with two different swarm sizes $N \in \{100, 150\}$ were performed and we could only reach diffusion rates in the interval $0.5 < 0.5B^2 < 0.9$. The flow decreases with the diffusion, as expected (see Figure 5.26 on page 111, error bars show 95% confidence intervals).

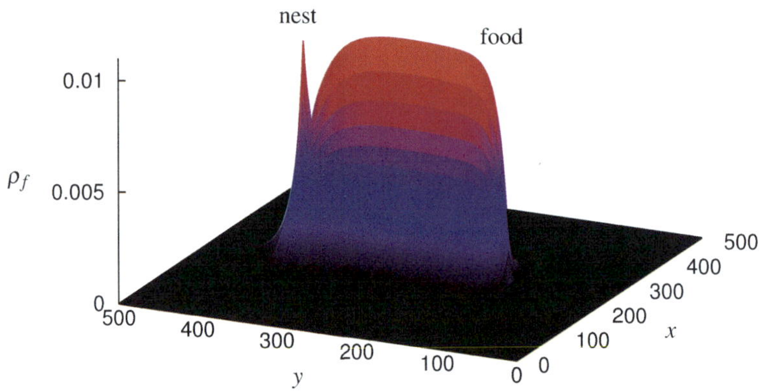

(a) Prediction of the model for the steady-state.

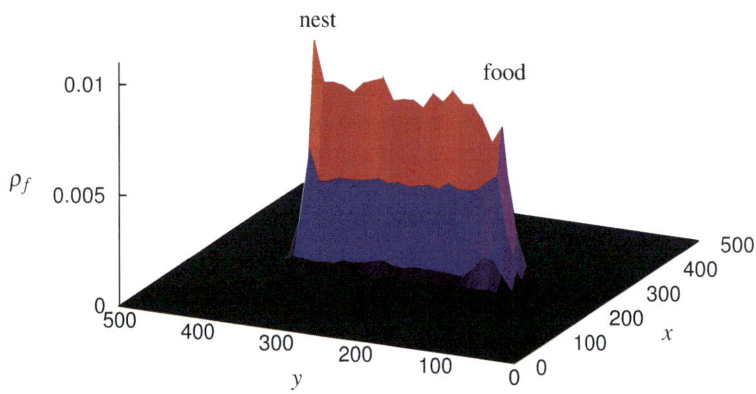

(b) Averaged over 30 simulation runs with $N = 100$ agents.

Fig. 5.25 Distribution of ρ_f for $0.5B^2 \approx 0.55$

In order to maximize the flow and following these results one would like to set the diffusion as low as possible. However, our observations of the simulation indicate that the lower the diffusion rate is the more unstable the system becomes. Please note, that the situation described here will only occur in simulations that implement interference effects between agents and an unbiased gradient ascent. If the robots follow the pheromone gradient $\frac{\partial P_f}{\partial r}$ very greedily, it becomes highly probable that they accumulate at certain spots. This might be caused by and lead itself to local maxima in the pheromone intensities P_f, where robots of state s_f are attracted. These groups of robots block others in state s_n traveling in the opposite direction. Hence, the local intensities P_f are reinforced and more and more robots accumulate at this spot. As a consequence, the flow of food might even break down temporarily

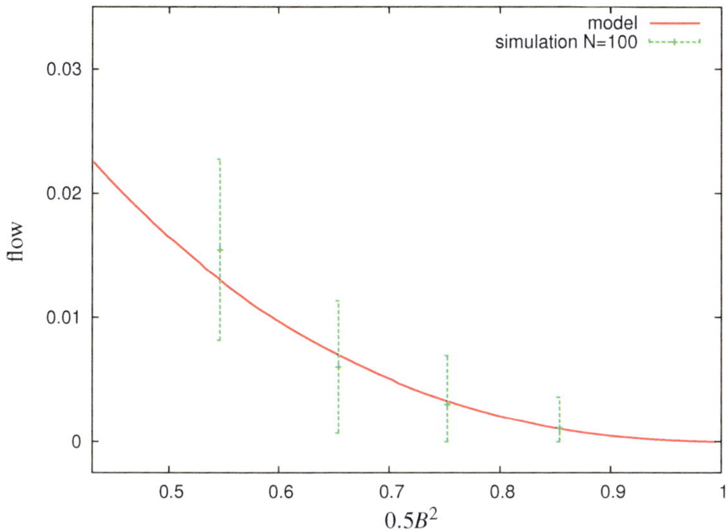

(a) 100 agents

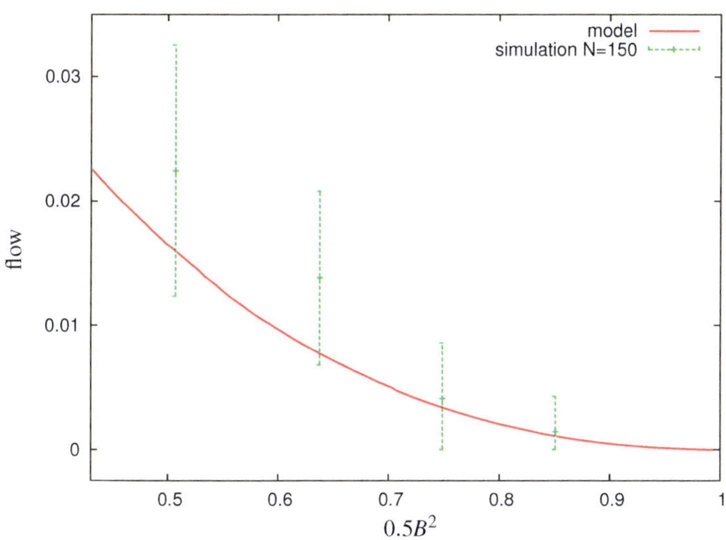

(b) 150 agents

Fig. 5.26 Average normalized flow of food as a function of the diffusion $0.5B^2$

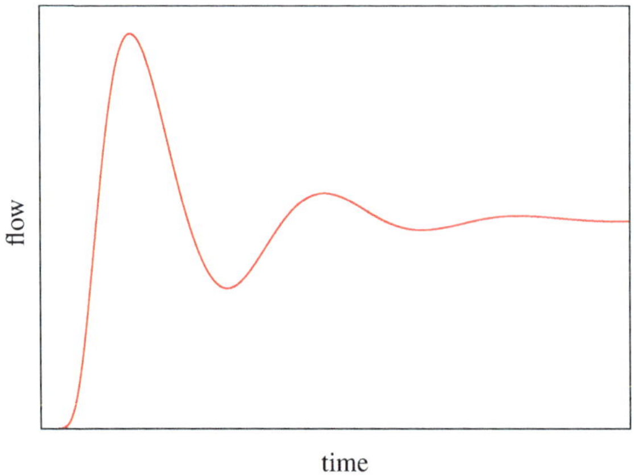

(a) Prediction by the model.

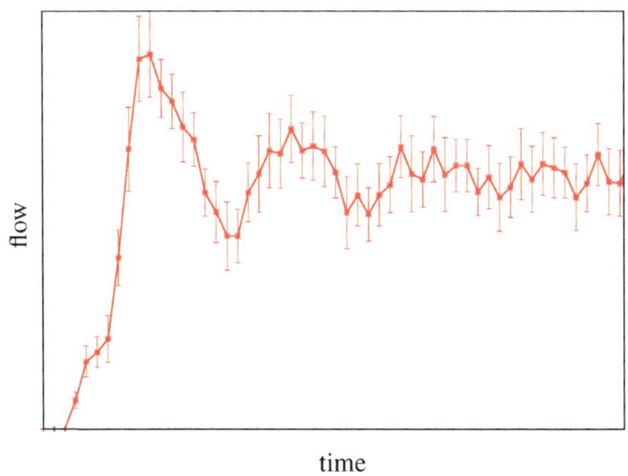

(b) Averaged over 30 simulation runs, error bars show 95% confidence interval.

Fig. 5.27 Qualitative comparison of the food flow predicted by the model and measured in the simulation

(cf. Figure 5.28 on the next page). This fact is not directly represented by the flow prediction of our model as it only gives the average flow of a functioning swarm without modeling such temporary breakdowns. While in the flow diagram of the model the intensity and number of oscillations only increase with decreasing diffusion, the batch-wise flow caused by robots moving in large groups and the temporary stop of the flow appears as an overshooting in the flow diagram of the model (see Figure 5.27).

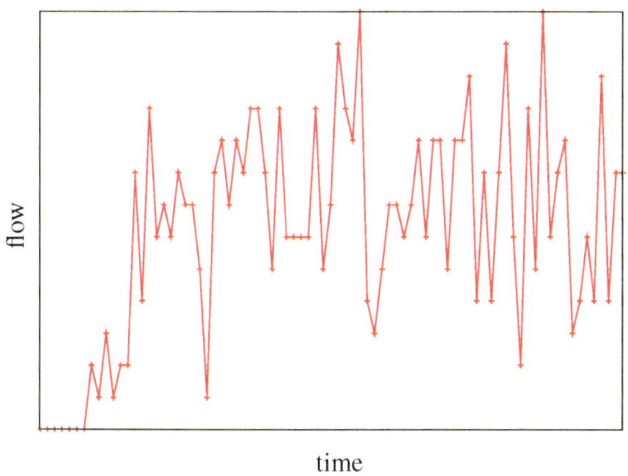

(a) Single simulation run showing stable behavior.

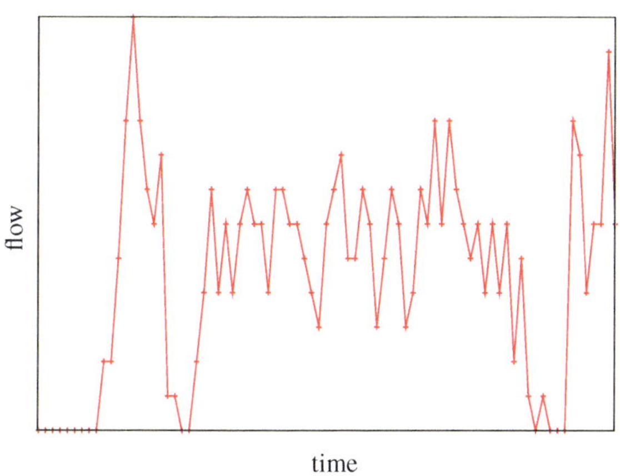

(b) Single simulation run showing unstable behavior.

Fig. 5.28 Qualitative comparison of the food flow of single simulation runs

5.5 Robot Aggregation as a Form of Computation

As discussed in Section 4.7 on page 68 the transient behavior of a robot swarm, that slowly adapts to a given environmental setting and eventually reaches a steady state until the environment changes again, can be interpreted as a form of computation. In the following we will investigate this hypothesis on the basis of an example.

We focus on the Euclidean Steiner tree (EST) problem which is basically the problem of connecting objects in a plane efficiently. The agent-based heuristic, that we investigated here, was shortly introduced in Hamann and Wörn (2007c) and presented in detail in Hamann and Wörn (2008a). The objects in the plane are connected by placing mobile relay stations, that we call robots in the following, instead of using wires. Starting with a uniformly distributed population of robots they aggregate in a way similar to diffusion-limited aggregation (DLA), see Tom A. Witten and Sander (1981). However, unlike DLA all robots are always moving and turn to avoid collisions between two moving robots. The objects to be connected serve as seeds from which trees of aggregated robots "grow". The use of such "random trees" for planning in robotics was introduced by LaValle and Kuffner (2001). In contrast to our approach these random trees are only virtual and are centrally controlled. As we are growing several trees or clusters at the same time this approach is also connected to diffusion-limited cluster–cluster aggregation (Kolb and Herrmann, 1985). However, the clusters in the present work are static.

5.5.1 The Euclidean Steiner Tree Problem

The EST problem is named after the Swiss mathematician Jakob Steiner and is defined as follows: A given set Z of N points or terminals in a plane has to be connected by lines of minimal length and, in contrast to the related minimal spanning tree problem, it is allowed to add a set of extra points S, called *Steiner points* (for an example see Figure 5.30(d) on page 117). The resulting network is a graph $G = (V,E)$ with nodes $V = Z \cup S$ and edges E accordingly defined. The probably better known instance of the Steiner tree problem class is defined on graphs where Steiner points can be picked from a finite set of points instead of placing them anywhere in a plane.

There are many applications to this problem in circuit design, mining, network design, and routing in ad hoc networks. Computing an optimal EST is NP-hard, that is, no efficient algorithm is known and is unlikely to be found, and the discretized variant is NP-complete (Garey et al., 1976). It is even hard to find an approximation within $95/94$ of the optimum (Chlebík and Chlebíková, 2002). A lot of work has been done to find both, better exact algorithms as well as polynomial time heuristics (Hwang et al., 1992; Robins and Zelikovsky, 2000). The best known heuristic was presented by Zachariasen and Winter (1999) that usually yields solutions close to the optimum within a few seconds at least for $N < 1000$. A software to compute exact EST is the GeoSteiner package (Warme et al., 2008).

An amusing anecdote reported by Aaronson (2005) gives a reason to suspect that anyway not too much intelligence might be of need for approximations of lower quality:

> Yet a well-known piece of computer science folklore maintains that, if two glass plates with pegs between them are dipped into soapy water, then the soap bubbles will rapidly form a Steiner tree connecting the pegs, this being the minimum-energy configuration.

Aaronson even experimented with real soap bubbles. He observed correct solutions but also cycles, incomplete trees, entirely different trees for the same configuration, and a relaxation process of several seconds leading to better soap bubble configurations. He follows that soap bubbles do not solve NP-complete problems in polynomial time. However, we note that (very) alternative approaches might be quite productive, too.

5.5.2 Growing Random Trees

We restrict ourselves to a two-dimensional setting but the proposed algorithm would work in the same way in three dimensions. We assume objects called *seeds* being placed in a bounded plane. Each seed represents a terminal out of Z of a considered Steiner tree problem. Furthermore we need some kind of robots that can move in the given environment. Whether these robots drive, crawl, fly, swim, or submerge is not relevant as long as they are able to move and to remain with sufficient accuracy at one spot. They have to be equipped with sensors allowing them to perceive other robots and seeds within a very short range compared to the dimensions of the bounded plane. Furthermore, they should be able to communicate to and measure an approximate bearing of other robots in their neighborhood. Their control mechanism suffices to be reactive. A large group of such mobile robots is positioned uniformly distributed over the whole plane. This could be done by the swarm autonomously in a previous phase.

In the following, we explain the control algorithm of the robots. See Figure 5.29 on the next page for a schematic overview. In general the robots move forward, try to find seeds, and listen for pings by robots being already connected to a seed. When a robot finds a seed (not shown in the schematic) and does not receive any pings it stops next to it. The robot generates a tree identification number (tree ID), that is with high probability unique, for example, by using a big random number. Then it starts to ping. When another robot receives this signal it checks back with the sender, if there is still an open slot (only three connections per robot are allowed). If so it maximizes the angles between itself, the sender, and the sender's neighbors. In case of one neighbor it forms a straight line. In case of two neighbors they try to reach a configuration with 120°, but only if this requires the movement of no more than two robots (local optimization). This can be achieved by some communication overhead and relying on relative angles only, see Hamann (2006) for details. This restriction of three connections per robot and the consequent angle of 120° is not arbitrarily chosen, as the optimal solution of a Steiner tree problem always consists of Steiner points of degree three and angles of 120° only. Additionally, we want to cover as much space in the plane as possible with a minimal number of stopped robots. The optimal solution to this tiling problem is provably the hexagon as it is found in honey-combs (Hales, 2001). However, the robots form only partial hexagons because the result should be a tree (a cycle-free graph).

Using this control algorithm the robots perform a process similar to diffusion-limited aggregation (Tom A. Witten and Sander, 1981). Provided with a sufficient number of robots and time a tree will grow at each seed. At some time, two trees will be connected. This is the case if a robot approaches two aggregated robots of

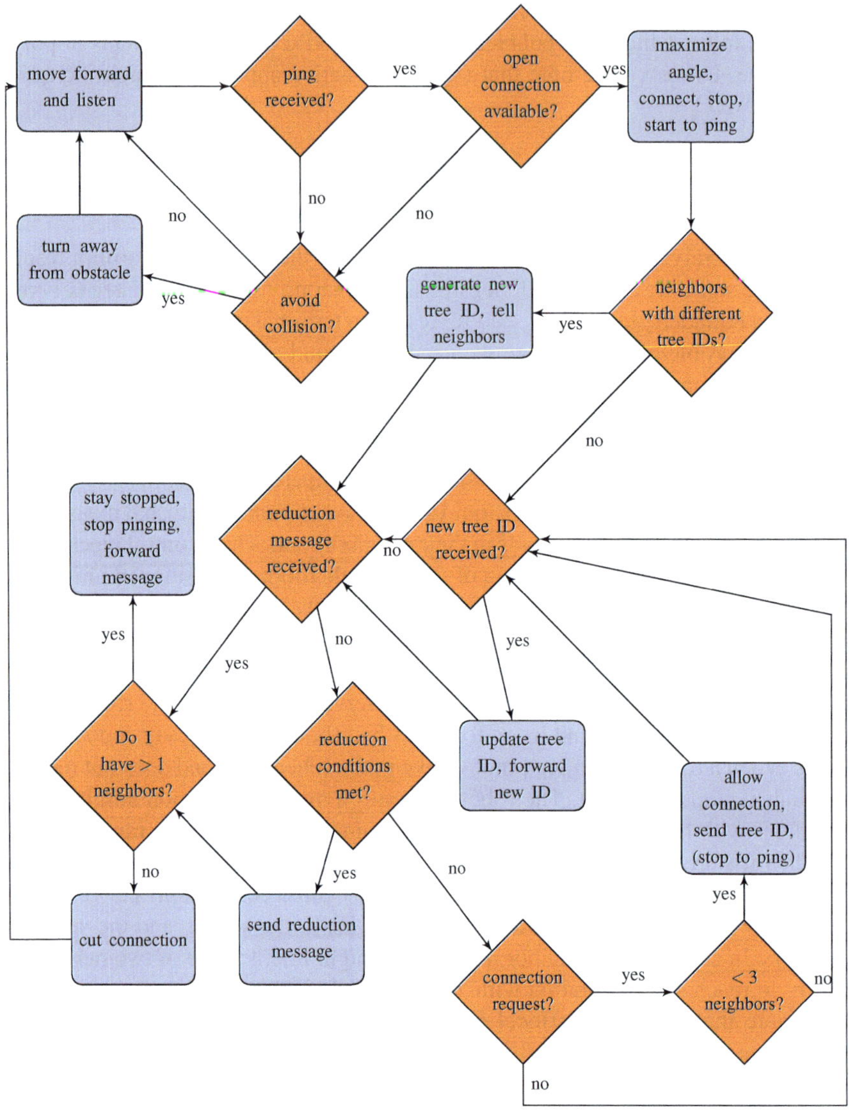

Fig. 5.29 Schema of the robot controller for the random tree growth algorithm

different trees virtually at the same time. By communicating their tree ID they ensure not to form a loop and agree upon a new tree ID, which is propagated through the new tree. Later, more trees will join. When only one huge tree is left, cf. Figure 5.30(a) on the facing page, or at an assigned time (reduction condition), a new process is started: All robots being connected to only one other robot, that is, they are leafs, will cut this connection and leave. In a chain reaction all unnecessary

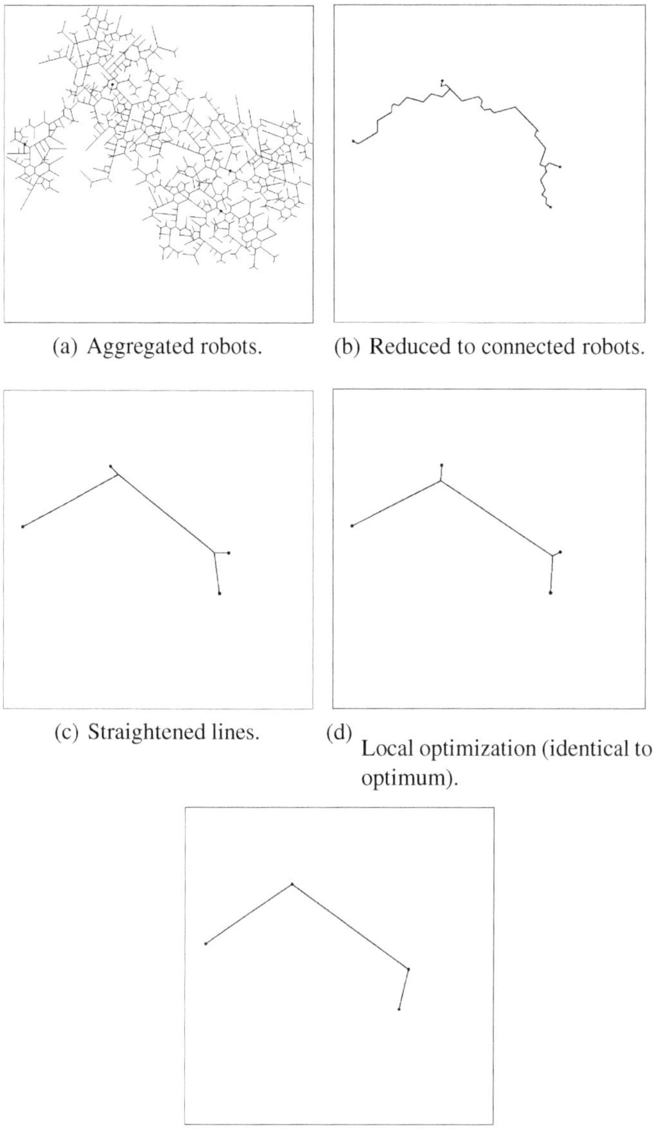

(a) Aggregated robots.

(b) Reduced to connected robots.

(c) Straightened lines.

(d) Local optimization (identical to optimum).

(e) Minimal spanning tree.

Fig. 5.30 Phases of the heuristic and minimal spanning tree

robots cut their connections and a tree consisting of a relatively small number of robots is left, cf. Figure 5.30(b). After this reduction robots being connected to three other robots represent Steiner points. By straightening the connections between the seeds and the Steiner points (i.e., releasing surplus robots in between, for details see Hamann (2006)), we get a first approximation to the optimal Steiner tree, cf.

Figure 5.30(c) on the preceding page. Note, the positions of the Steiner points were determined dynamically during the tree growth process. Interpreting this in the sense of swarm intelligence: the Steiner point set S is the result of a collective decision that emerges from the numerous agent–agent interactions. An additional improvement is achieved by locally rearranging the Steiner points towards their optimal position (an approximation to the Fermat–Torricelli point). This is achieved by moving the robots at the Steiner point towards the direction of the smallest angle. This might result in the optimal solution as shown in Figure 5.30(d) on the page before but in general the optimal configuration of the Steiner points is only achievable by global optimization.

Keeping a reasonable amount of redundant robots in the lines this heuristic is robust to breakdowns of single robots although it might seem very inefficient. In addition, it is scalable because of its totally local approach. Whether this method can be a fast way of approximating a Steiner tree is a question of the reaction times and speed of the robots. At the time mass production of such devices will become possible, this scenario might actually be feasible, see Section 2.3.2 on page 12.

5.5.3 Results

The following results were obtained using a first-order geometric simulation with continuous space. Our emphasis is on the general behavior of the agent system which we claim to be covered by this kind of simulation. A more complex simulation would have multiplied the computational complexity. However, already this abstract simulation kept a personal computer busy for days computing thousands of runs using 10^5 and more robots. As we do not yet have such quantities of robots or other computational devices available we had to simulate our massively parallel heuristic serially.

We compare the results of our heuristic to the optimal solution and the minimal spanning tree — the typical benchmark problem for Steiner tree heuristics. For this purpose we compare the reduction in percent r of the (suboptimal) Steiner tree length L_{steiner} to the minimal spanning tree length L_{spanning}

$$r = \frac{L_{\text{spanning}} - L_{\text{steiner}}}{L_{\text{spanning}}} \cdot 100\% \tag{5.38}$$

with

$$L = \sum_{(\mu,\nu)\in E} \|\mu - \nu\|_2, \tag{5.39}$$

$\mu,\nu \in \mathbb{R}^2$ are node positions and $\|\cdot\|_2$ denotes the Euclidean norm.

For the optimal Steiner tree this value ranges from instance to instance between $r_{\min} = 0\%$ (minimal Steiner and spanning tree are identical) and $r_{\max} > 10\%$. For the minimal Steiner tree the reduction averaged over many instances converges almost independently of the terminal number N to $\bar{r}_{\text{opt}} \approx 3.1\%$.

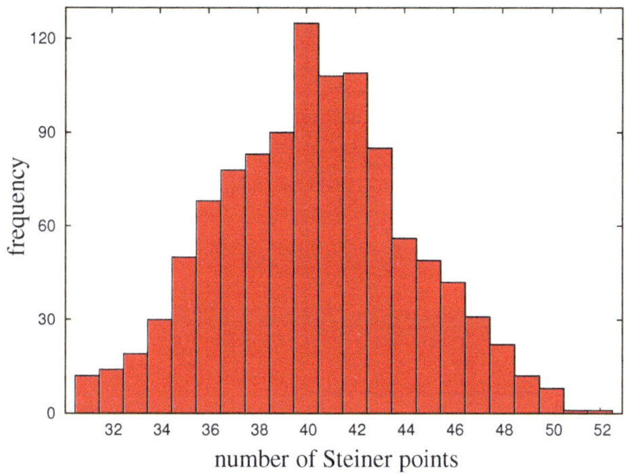

Fig. 5.31 Histogram of Steiner points generated in 550 samples by the heuristic for a problem instance of 100 terminals showing a mean of about 40.2, optimal are 38 Steiner points, and maximally possible are 98.

Before we compare the actual performance we have a look at the number of Steiner points generated by the random tree heuristic. In principal this number ranges from 0 (spanning tree) to $N-2$. However, it turns out that the heuristic generates reasonable numbers of Steiner points. See Figure 5.31 showing the situation of a single problem instance with a mean that overestimates the optimal number of Steiner points by 5.8%.

Now we compare our heuristic to the exact solution focusing on the governing parameter, the robot number. We omit a time analysis since the time consumption can be kept constant with increasing robot number (due to strictly local actions of each robot). Only the serialized simulation of the heuristic suffered from the complexity of high robot numbers. The solutions become better the more robots are used as shown in Figure 5.32(a) on the following page. The reduction decreases linearly in problem size. Due to the probabilistic characteristic of the proposed heuristic the average performance is improved by repeated runs. In Figure 5.32(b) we show the performance achieved by selecting the best solution out of 50 runs. Especially for bigger instances $N > 50$ the performance could be furthermore improved by increasing the number of robots.

While our heuristic is inferior to the state-of-the-art heuristics in the performance we identify its advantage in its adaptivity and due to the decentralized approach also in its robustness. We test the adaptivity by replacing a terminal after 40 time steps. The terminal configuration shown in Figure 5.33 on page 121 and 10^4 robots were used. Only 8.2% were irregular approximations (not all terminals connected) evaluating 500 samples. The average reduction dropped by 42% compared to the heuristic started with the final terminal configuration.

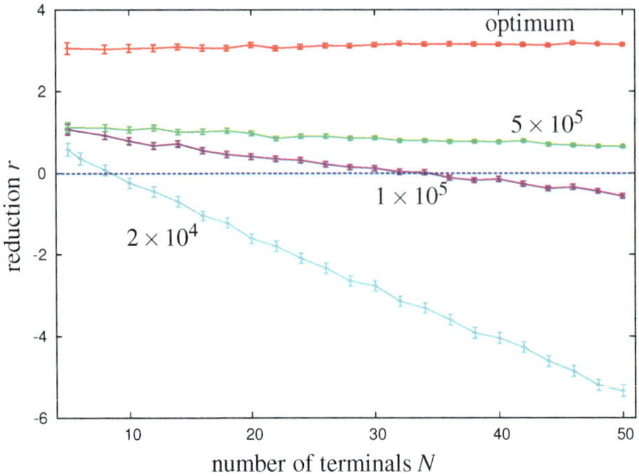

(a) 2×10^4, 1×10^5, and 5×10^5 robots, 500 samples for each point.

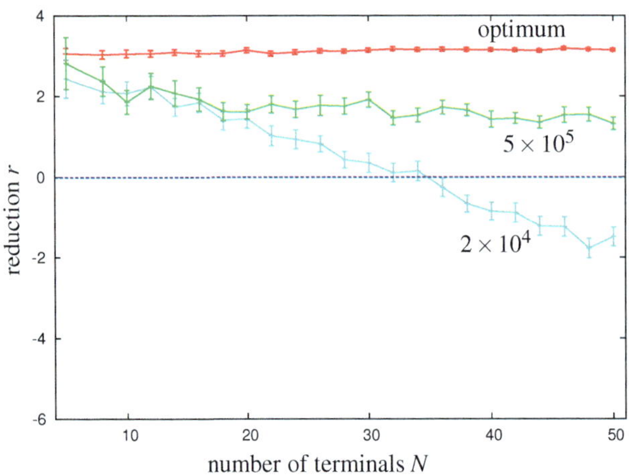

(b) Taking only the best out of 50 runs using 2×10^4 (80 samples per point) and 5×10^5 (60 samples per point) robots.

Fig. 5.32 Comparing the reduction of the heuristic for different robot numbers and varied problem size N to the optimal solution; error bars are 95% confidence intervals

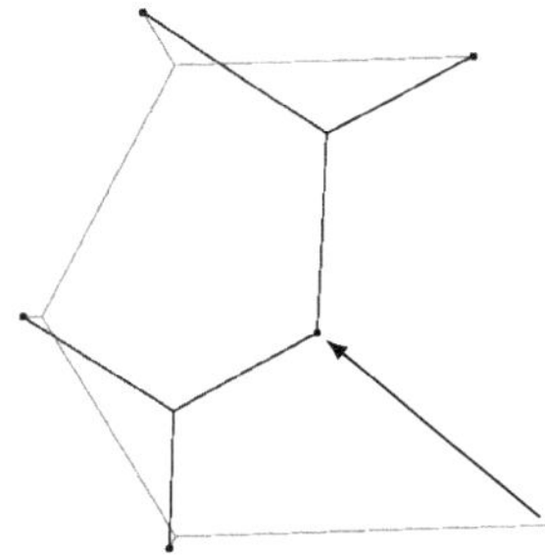

Fig. 5.33 Superimposed optimal solutions for the beginning set of terminals (gray) and the final set of the adaptivity test scenario. The arrow indicates the rearranged terminal

5.5.4 Discussion

The heuristic proposed in this paper is definitely not superior to the state-of-the-art heuristics concerning quality and computing time (the best known heuristic typically delivers approximations within about 4% from optimum (Zachariasen and Winter, 1999)). However, it shows a reasonable degree of adaptivity. Furthermore, it is supposed to be quite robust as there is no single point of failure. Comparing the fault tolerance of our method to the classical approach corresponds to answering the question: How wrong can wrong be? One might argue that this comparison is unfair because the difference is only due to different output methods: explicit and physical (motion, positions, angles) compared to symbolic (numbers, calculations). However, this actually is the fundamental difference in the method of information processing between these approaches. The classical computer processes abstract symbols while the agent system processes physical positions. Therefore, we consider a comparison to be fair and the difference in the wrongness is big. A single error might cause almost infinitely high deviations using a symbol-based approach as there are virtually no limitations for what could happen to a symbol in the CPU or the memory. For example, a single bit shift might lead to negative distances. This is in contrast to our agent system, where a single error might break two subtrees but arbitrary deviations are impossible. This is one advantage of the strictly bounded operating range of the robots, limiting not only their possibilities but also the consequences of errors. An analysis of the proposed algorithm would obviously be very hard due to its properties that might be considered "emergent". For example, the growth of a random tree is dependent on its relative position to other trees. Any model describing the tree growth independently would have little explanatory power. Including all trees into the model would increase its complexity significantly.

5.6 Summary

In this chapter we have applied the model framework to several scenarios including experiments with real robots and with simulated robots for the purpose of a validation. In the case of the collision-based adaptive swarm aggregation we were able to show that our model is able to predict the dynamics and the spatial features of a real robotic swarm with satisfying accuracy. The model captures the relevant macroscopic features that define the swarm performance while abstracting distracting microscopic details.

In the collective perception scenario we have shown that a scenario based on simple "one-hop" communication can be modeled as well although several parameters needed to be determined empirically. However, this is the challenge of modeling communication in closed-form as discussed in Section 4.5.2 on page 66 and reflects the state-of-the-art of the theory of ad hoc and sensor networks.

In the case of the emergent taxis scenario we have shown that a qualitative model can be obtained quite fast by simply applying one or two key abstractions. While quantitative models of high accuracy are desirable a qualitative model obtained in a short development phase is helpful as well and can lead to major insights in the research of swarms.

The work on the foraging scenario has shown the application of the model to the case of several states with a probability distribution for each state.

The scenario of robot aggregation as a heuristic for hard problems has shown that our concept of interpreting swarm dynamics as computation is not just an abstract theory. We have shown that a swarm is qualified to perform computations.

The variety of investigated scenarios shows also the potentially vast domain of applications of this model. This is, for example, indicated by the different approaches of adapting function **A** to the scenarios. In the collision-based scenario we were able to minimize the framework model to drop **A** because the deterministic components of the robot behavior lead to homogenized robot densities. In the collective perception scenario the **A**-function represented the directional information a robot obtains by means of communication and the establishment of a virtual potential field. In the emergent taxis scenario the **A**-function was used to incorporate an average behavior, which is, however, spatially inhomogeneous and dynamic. Finally, in the foraging scenario **A** was defined by a gradient and the exploitation rate.

Chapter 6
Conclusion and Outlook

6.1 Conclusion

In this work, we have presented a space- and time-continuous, generic model framework of communicating robot swarms. This model is distinguished by its strong connection between the micro- and the macro-level and the variety of potential applications. The model was, in a validation step, successfully applied to a variety of sample scenarios proving the general feasibility of the proposed approach, the models universalism, and its accuracy.

In addition, we have proposed the interpretation of swarm dynamics as a form of computation. This was also exemplified in an application to a hard problem. The research field of swarm robotics, to which this work belongs, was fitted into the scientific literature by giving an extensive overview of related concepts with an emphasis on swarm models. Thus, this work as a whole introduces the reader into the field of swarm control algorithm design including related far-reaching questions and providing all that is needed to begin the development of swarm models.

The proposed model framework is an advancement in several respects compared to previous approaches:

- The model establishes (at least a partial) connection between the microscopic level (Langevin equation) and the macroscopic level (Fokker–Planck equation).
- A successful application of methods of statistical mechanics to robot swarms was neither shown before nor was it expected.
- Space is continuously represented in the model. Thus, inhomogeneous space (e.g., inhomogeneous density distributions) is presentable.
- The model is written down in a concise form in contrast to, for example, simulations.
- There is a vast variety of scenarios to which the model can be applied.

Furthermore, the noteworthiness of this work is the diversity of its methodology as it spans such divers fields as differential equations, biology, and robotics.

H. Hamann: Space-Time Cont. Models of Swarm Rob. Sys., COSMOS 9, pp. 123–125.
springerlink.com

6.2 Outlook

A research project, as the work at hand, raises new questions, most likely more than it answers. This is especially true for a work in a very new field of research such as swarm robotics. For the proposed model the hope seems to be legitimate that it will be of use in both application and theory. This is due to its universalism and its theoretical soundness in contrast to phenomenological approaches. However, there are many possibilities of extensions of the proposed model and of completely new approaches.

A first extension could be the incorporation of statistical values. Already the knowledge of the variance for the mean density at a given point in space would be extremely helpful. A broad extension would be to hold a probability density in each grid point although this would introduce a high computational complexity.

In a next step the model state might be needed to be extended by a representation of velocities or directions in order to allow the Markov property for more deterministic scenarios and to reach higher accuracy. However, this will lead to a much more complex formalism increasing the complete overhead of deriving, maintaining, and adapting the model.

For rather deterministic control algorithms as in the case of the collision-based aggregation (see Section 5.1 on page 77) an approach using something similar to the Lévy walk might prove to be more successful because the robots move frequently untouched through the middle third of the arena.

A challenging question is the reported imbalance between the Langevin and the Fokker–Planck equation. While certain abstractions are easily possible using the Fokker–Planck equation, these abstractions are difficult in case of the Langevin equation (see Section 4.8.1 on page 72). This might be a difference between models of trajectories and models of ensembles as a matter of principle.

The challenge of modeling communication is also an open question. Whether a further effort towards an analytical solution of this problem is useful, is questionable. The combination of the proposed model with sampling from a microscopic model might be the more advisable way.

This raises the complex question whether the space-centered approach (model based on points in space) is superior to the agent-centered (model based on data related to individual agents) approach. A question that could be investigated using the proposed framework.

The primary objective of this work was to develop an efficient way of designing a model based on a given control algorithm. In order to increase the efficiency of the algorithm design phase essentially, it is important to draw conclusions that allow improvements of the current algorithm. These conclusions should be as precise as possible. Using the proposed framework this is done by defining an appropriate performance metric followed by parameter scans. The subject of a future research project could, thus, be to develop methods to trace necessary changes of significant features in the current algorithm design.

Further possible research issues, that might prove to be fruitful, include particular applications of the proposed model. For example, it could be used to do an in-depth comparison of biologically inspired to classical algorithms – an enterprise that is overdue. Another possibility is to use the model in an online approach. While the main purpose of this work was to produce a tool for helpful offline predictions (i.e., before the system is implemented) it could still also be used to give short term predictions online. Periodic measurements could be used to calibrate the model parameters allowing predictions of higher accuracy.

Further validation of the model and the concept of embodied computation is in need, too.

6.3 Acknowledgments

At first I want to thank Prof. Dr. Heinz Wörn and Prof. Dr. Rüdiger Dillmann for their support.

Many thanks to my colleagues who helped me by discussing my ideas and by giving advice: Thomas Schmickl, Ramon Estaña, Sebastian Schuster, Sergey Kornienko, Marc Szymanski, and Andreas Schmid. I also thank the many anonymous reviewers that reviewed my papers and often gave great remarks or advised me of some overlooked publication.

Thanks to the developers of the free software emacs, gcc, gnuplot, LaTeX, pgf and tikz, MediaWiki, and Firefox which were of great help in my daily work.

During his time at the Karlsruhe Institute of Technology (formerly known as University of Karlsruhe), the author was supported by the German Research Foundation (DFG) within the Research Training Group GRK 1194 "Self-organizing Sensor-Actuator Networks".

Appendix A
Numerical Simulation of the Langevin Equation

```
#define A 0.1
#define B 0.3
#define TAU 0.1
#define ITERATIONS 500
#define SEED 1
#define Pi 3.141592654

#include <stdio.h>
#include <stdlib.h>
#include <time.h>
#include <math.h>

// discrete uniform distribution:
float discreteUniform() {
  if(((float)rand()/(float)RAND_MAX)<0.5)
    return -1.0;
  else
    return 1.0;
}

// continuous uniform distribution:
float continuousUniform() {
  return (2.0*((float)rand()/(float)RAND_MAX)-1.0);
}

// normal distribution (using the Box-Muller method):
float normal() {

  float u1, u2;

  u1 = (float)rand()/(float)RAND_MAX;
  u2 = (float)rand()/(float)RAND_MAX;
  return sqrt(-2.0*logf(u1))*cos(2.0*Pi*u2);
}
```

```
int main(int argc, char **argv) {

  FILE *f;
  float dx;
  float x=0.0;
  int count=0;

  srand(SEED);
  f = fopen("trajectory","w");
  while((x<100.0) && (x>-100.0) && (count < ITERATIONS)) {

    // dx = continuousUniform();
    // dx = discreteUniform();
    dx = normal();
    x += dx*sqrt(TAU)*B;
    x += A*TAU;
    count++;
    fprintf(f, "%e %e\n", (float)count*TAU, x);
  }
  fclose(f);
  return 0;
}
```

Appendix B
Simple Numerical Solver of the Fokker–Planck Equation

```
#define GRID_LENGTH   100
#define ITERATIONS  10000
#define TAU 0.001
#define A 0.1
#define B 0.3
#define Q 1.0

#include <math.h>
#include <stdio.h>

double update(int x, double *grid, double *potential) {

  double star, drift;

  star  = B*B * (grid[x-1] - grid[x]);
  star += B*B * (grid[x+1] - grid[x]);

  if(potential[x-1] - potential[x] < 0.0)
    drift = -A * (grid[x-1] - grid[x]);
  else
    drift = A * (grid[x-1] - grid[x]);

  return grid[x]
    + TAU * (0.5 * Q * star + drift);
}


int main(int argc, char **argv) {

  int i, x;
  double grid_now[GRID_LENGTH];
  double grid_next[GRID_LENGTH];
  double potential[GRID_LENGTH];
  double *temp;
  FILE *f;
```

```
  init_grid(grid_now);
  init_potential(potential);

  for(i=0; i < ITERATIONS; i++) {

    for (x=0; x < GRID_LENGTH; x++)
      grid_next[x] = update(x, grid_now, potential);

    temp = grid_now;
    grid_now = grid_next;
    grid_next = temp;

  }

  f = fopen("grid","w");
  for (x=0; x < GRID_LENGTH; x++)
    fprintf(f, "%d %e\n", x, grid_now[x]);
  fclose(f);

  return 0;

}
```

References

Aaronson, S.: NP-complete problems and physical reality. ACM SIGACT News 36(1), 30–52 (2005) Cited on p. 114.

Abbott, R.: Emergence, entities, entropy, and binding forces. In: The Agent 2004 Conference on: Social Dynamics: Interaction, Reflexivity, and Emergence, Argonne National Labs and University of Chicago (October 2004) Cited on p. 8.

Abbott, R.: Emergence explained: getting epiphenomena to do real work. arXiv:cs/0602045 (February 2006) Cited on p. 35.

Abelson, H., Allen, D., Coore, D., Hanson, C., Homsy, G., Knight, T., Nagpal, R., Rauch, E., Sussman, G., Weiss, R.: Amorphous computing. Communications of the ACM 43(5), 74–82 (2000) Cited on p. 17.

Alexander, J.C., Giesen, B., Münch, R., Smelser, N.J. (eds.): The Micro-Macro Link. University of California Press, California (1987) Cited on p. 9.

Anderson, P.W.: More is different. Science 177(4047), 393–396, 24 (1972) Cited on pp. 21 and 23.

Arkin, R.C.: Motor schema based mobile robot navigation. International Journal of Robotics Research 8 (1989) Cited on p. 105.

Barthelemy, P., Bertolotti, J., Wiersma, D.S.: A Lévy flight for light. Nature 453, 495–498 (2008) Cited on p. 88.

Bayindir, L., Şahin, E.: A review of studies in swarm robotics. Turkish Journal of Electrical Engineering and Computer Sciences 15, 115–147 (2007) Cited on p. 8.

Beal, J., Bachrach, J.: Infrastructure for engineered emergence on sensor/actuator networks. IEEE Intelligent Systems 21(2), 10–19 (2006) Cited on p. 38.

Bedau, M.A.: Weak emergence. Philosophical Perspectives: Mind, Causation, and World 11, 375–399 (1997) Cited on pp. 22 and 23.

Beni, G.: From swarm intelligence to swarm robotics. In: Şahin, E., Spears, W.M. (eds.) Swarm Robotics – SAB 2004 International Workshop, Santa Monica, CA. LNCS, July 2005, pp. 1–9 (2005) Cited on pp. 8 and 12.

Beni, G., Wang, J.: Swarm intelligence in cellular robotic systems. In: Proceedings of the NATO Advanced Workshop on Robots and Biological Systems (1989) Cited on p. 9.

Bettstetter, C.: On the connectivity of Ad Hoc networks. The Computer Journal 47(4), 432–447 (2004) Cited on p. 66.

Bickhard, M.H., Campbell, D.T.: Emergence. In: Andersen, P.B., Emmeche, C., Finnemann, N.O., Christiansen, P.V., Voetmann, P. (eds.) Downward Causation: Minds, Bodies and Matter, pp. 322–348. Århus University Press (2000) Cited on p. 24.

Bishop, R.C.: Nonequilibrium statistical mechanics brussels-austin style. Studies In History and Philosophy of Science Part B: Studies In History and Philosophy of Modern Physics 35(1), 1–30 (2004) Cited on p. 30.

Bjerknes, J.D., Winfield, A., Melhuish, C.: An analysis of emergent taxis in a wireless connected swarm of mobile robots. In: IEEE Swarm Intelligence Symposium, pp. 45–52. IEEE Press, Los Alamitos (2007) Cited on pp. 8, 15, 29, 100, 101, and 105.

Bonabeau, E.: Predicting the unpredictable. Harvard Business Review 80(3) (March 2002) Cited on pp. 8 and 36.

Bonabeau, E., Thraulaz, G., Fourcassi, V., Deneubourg, J.-L.: The phase-ordering kinetics of cemetery organization in ants. Technical Report 98-01008, Santa Fe Institute (1998) Cited on p. 32.

Bonabeau, E., Dorigo, M., Theraulaz, G.: Swarm Intelligence: From Natural to Artificial Systems. Oxford Univ. Press, Oxford (1999)

Braitenberg, V.: Vehicles: Experiments in synthetic psychology. MIT Press, Cambridge (1984) Cited on pp. 9 and 70.

Broad, C.D.: The Mind and Its Place in Nature. Kegan Paul, Trench, Trubner & Co., London (1925) Cited on p. 17.

Brown, R.: A brief account of microscopical observations made in the months of June, July and August, 1827, on the particles contained in the pollen of plants; and on the general existence of active molecules in organic and inorganic bodies. Philosophical Magazine 4, 161–173 (1828) Cited on p. 21.
Cited on p. 42.

Brufau, J., Puig-Vidal, M., López-Sánchez, J., Driesen, W., Breguet, J.-M., Snis, N., Simu, U., Johansson, S., Gao, J., Velten, T., Seyfried, J., Estaña, R., Wörn, H.: Micron: Small autonomous robot for cell manipulation applications. In: Proc. of the 2005 IEEE Int. Conf. on Robotics and Automation (ICRA), pp. 856–861. IEEE Press, Los Alamitos (2005) Cited on p. 12.

Camazine, S., Deneubourg, J.-L., Franks, N.R., Sneyd, J., Theraulaz, G., Bonabeau, E.: Self-Organizing Biological Systems. Princeton Univ. Press, Princeton (2001) Cited on pp. 7 and 10.

Caprari, G., Balmer, P., Piguet, R., Siegwart, R.: The autonomous microbot 'alice': a platform for scientific and commercial applications. In: Proc. of the Ninth Int. Symp. on Micromechatronics and Human Science, Nagoya, Japan, pp. 231–235 (1998) Cited on pp. 15 and 63.

de Castro, L.N.: Fundamentals of natural computing: an overview. Physics of Life Reviews 4(1), 1–36 (2007) Cited on p. 25.

Chlebík, M., Chlebíková, J.: Approximation hardness of the Steiner Tree problem on graphs. In: Penttonen, M., Schmidt, E.M. (eds.) SWAT 2002. LNCS, vol. 2368, pp. 170–179. Springer, Heidelberg (2002) Cited on p. 114.

Churchland, P.: Poetry in motion. Nature 450(7166), 29 (2007) Cited on p. 24.

Conte, R., Edmonds, B., Moss, S., Sawyer, R.K.: Sociology and social theory in agent based social simulation: A symposium. Computational & Mathematical Organization Theory 7(3), 183–205 (2001) Cited on p. 17.

Correll, N.: Coordination schemes for distributed boundary coverage with a swarm of miniature robots: synthesis, analysis and experimental validation. PhD thesis, Ecole Polytechnique Fédérale de Lausanne (2007) Cited on pp. 15 and 28.

Correll, N., Martinoli, A.: System identification of self-organizing robotic swarms. In: Gini, M., Voyles, R. (eds.) Proceedings of the 8th Int. Symp. on Distributed Autonomous Robotic Systems (DARS 2006), Berlin, Germany, pp. 31–40. Springer, Heidelberg (2006) Cited on p. 28.

Correll, N., Cianci, C., Raemy, X., Martinoli, A.: Self-organized embedded sensor/actuator networks for "smart" turbines. In: IEEE/RSJ International Conference on Intelligent Robots and Systems Workshop on Network Robot System: Toward intelligent robotic systems integrated with environments, IEEE Press, Los Alamitos (2006) Cited on pp. 15 and 16.

Crank, J., Nicolson, P.: A practical method for numerical evaluation of solutions of partial differential equations of the heat conduction type. Proceedings of the Cambridge Philosophical Society 43, 50–64 (1947) Cited on p. 55.

Crutchfield, J.: The calculi of emergence: Computation, dynamics, and induction. Physica D 75(1-3), 11–54 (1994a) Cited on p. 24.

Crutchfield, J.: Is anything ever new? In: Cowan, G., Pines, D., Melzner, D. (eds.) Complexity: Metaphors, Models, and Reality. SFI Series in the Sciences of Complexity proceedings, vol. 19, pp. 479–497. Addison-Wesley, Reading (1994b) Cited on p. 24.

Şahin, E.: Swarm robotics: From sources of inspiration to domains of application. In: Şahin, E., Spears, W.M. (eds.) Swarm Robotics 2004. LNCS, vol. 3342, pp. 10–20. Springer, Heidelberg (2005) Cited on pp. 8 and 14.

Şahin, E., Spears, W.M. (eds.): Swarm Robotics 2004. LNCS, vol. 3342. Springer, Heidelberg (2005) Cited on pp. 1 and 12.

Darley, V.: Emergent phenomena and complexity. In: Brooks, R., Maes, P. (eds.) Artificial Life IV, pp. 411–416 (1994) Cited on p. 8.

Deguet, J., Demazeau, Y., Magnin, L.: Elements about the emergence issue: A survey of emergence definitions. Complexus 3(1-3), 24–31 (2006) Cited on pp. 8 and 36.

Deneubourg, J.-L., Aron, S., Goss, S., Pasteels, J.M.: The self-organizing exploratory pattern of the argentine ant. Journal of Insect Behavior 3, 159–168 (1990) Cited on pp. 7 and 32.

Doob, J.L.: Stochastic Processes. Wiley, New York (1953) Cited on p. 47.

Dorigo, M., Di Caro, G.: Ant Colony Optimization: A new meta-heuristic. In: Angeline, P.J., Michalewicz, Z., Schoenauer, M., Yao, X., Zalzala, A. (eds.) Proceedings of the 1999 Congress on Evolutionary Computation (CEC 1999), Piscataway, NJ, pp. 1470–1477. IEEE Press, Los Alamitos (1999) Cited on p. 10.

Dorigo, M., Şahin, E.: Guest editorial: Swarm robotics. Autonomous Robots 17(2-3), 111–113 (2004) Cited on pp. 8 and 14.

Dorigo, M., Bonabeau, E., Theraulaz, G.: Ant algorithms and stigmergy. Future Generation Computer Systems 16(9), 851–871 (2000) Cited on p. 8.

Dorigo, M., Trianni, V., Şahin, E., Groß, R., Labella, T.H., Baldassarre, G., Nolfi, S., Deneubourg, J.-L., Mondada, F., Floreano, D., Gambardella, L.M.: Evolving self-organizing behaviors for a swarm-bot. Autonomous Robots 17(2-3), 223–245 (2004) Cited on p. 39.

Dorigo, M., Tuci, E., Trianni, V., Groß, R., Nouyan, S., Ampatzis, C., Labella, T.H., O'Grady, R., Bonani, M., Mondada, F.: Swarm-bot: Design and implementation of colonies of self-assembling robots. In: Yen, G.Y., Fogel, D.B. (eds.) Computational Intelligence: Principles and Practice, pp. 103–135. IEEE Press, Los Alamitos (2006) Cited on p. 15.

Edelstein-Keshet, L.: Mathematical models of swarming and social aggregation. Robotica 24(3), 315–324 (2006) Cited on p. 32.

Eigen, M.: The Hypercycle: A Principle of Natural Self Organization. Springer, Heidelberg (1979) Cited on p. 32.

Einstein, A.: Über die von der molekularkinetischen Theorie der Wärme geforderte Bewegung von in ruhenden Flüssigkeiten suspendierten Teilchen. Annalen der Physik 17, 549–560 (1905) Cited on p. 42.
Einstein, A., Penrose, R., Stachel, J.: Einstein's Miraculous Year: Five Papers That Changed the Face of Physics. Princeton University Press, Princeton (1998) Cited on p. 42.
ePuck. e-puck desktop mobile robot (2008), `http://www.e-puck.org/` Cited on p. 15.
Fahlbusch, S., Fatikow, S., Seyfried, J., Bürkle, A.: Flexible microrobotic system miniman: Design, actuation principle and control. In: Proc. of the 1999 IEEE/ASME International Conference on Advanced Intelligent Mechatronics (AIM 1999), Atlanta, GA, USA. IEEE Press, Los Alamitos (1999) Cited on p. 12.
Fatikow, S., Rembold, U.: Principles of micro actuators and their applications. In: Proc. of the IAPR Workshop on Micromachine Technologies and Systems (MTS), Tokyo, Japan, October 1993, pp. 108–117 (1993) Cited on p. 11.
Fatikow, S., Seyfried, J., Fahlbusch, S., Buerkle, A., Schmoeckel, F.: A flexible microrobot-based microassembly station. Journal of Intelligent and Robotic Systems 27(1-2), 135–169 (2000) Cited on p. 11.
Feddema, J.T., Lewis, C., Schoenwald, D.A.: Decentralized control of cooperative robotic vehicles: theory and application. IEEE Transactions on Robotics and Automation 18(5), 852–864 (2002) Cited on p. 28.
Fisher, R.A.: The genetical theory of natural selection. Oxford University Press, Oxford (1930) Cited on p. 32.
Fokker, A.D.: Die mittlere Energie rotierender elektrischer Dipole im Strahlungsfeld. Annalen der Physik 348(5), 810–820 (1914) Cited on p. 42.
Galstyan, A., Hogg, T., Lerman, K.: Modeling and mathematical analysis of swarms of microscopic robots. In: Proceedings of IEEE Swarm Intelligence Symposium (SIS-2005), Pasadena, CA, June 2005, pp. 201–208. IEEE Press, Los Alamitos (2005) Cited on p. 28.
Gardiner, C.W.: Handbook of Stochastic Methods for Physics, Chemistry and the Natural Sciences. Springer, Heidelberg (1985) Cited on p. 71.
Garey, M.R., Graham, R.L., Johnson, D.S.: Some NP-complete geometric problems. In: Annual ACM Symposium on Theory of Computing, pp. 10–22 (1976) Cited on p. 114.
Gazi, V., Passino, K.M.: Stability analysis of swarms. IEEE Transactions on Automatic Control 48(4), 692–697 (2003) Cited on p. 28.
Ginzburg, V.L., Landau, L.D.: Concerning the theory of superconductivity. Zurnal Çeksperimental'noj i Teoreticeskoj Fiziki 20(12), 1064–1082 (1950) Cited on p. 19.
Grassé, P.-P.: La reconstruction du nid et les coordinations interindividuelles chez bellicositermes natalensis et cubitermes sp. la théorie de la stigmergie:essai d'interprétation du comportement des termites constructeurs. Insectes Sociaux 6, 41–83 (1959) Cited on p. 9.
Grassé, P.-P.: Nouvelles experiences sur le termite de müller (macrotermes mülleri) et considerations sur la théorie de la stigmergie. Insectes Sociaux 14, 73–102 (1967) Cited on p. 9.
Grimmett, G.: Percolation. Grundlehren der mathematischen Wissenschaften, vol. 321. Springer, Berlin (1999) Cited on p. 7.
Grünbaum, D., Okubo, A.: Modeling social animal aggregations. Frontiers in Theoretical Biology 100, 296–325 (1994) Cited on p. 32.
Haken, H.: Synergetik – Die Lehre vom Zusammenwirken. Umschau in Wissenschaft und Technik 6, 191–195 (1971) Cited on pp. 7, 18, and 19.
Haken, H.: Synergetics – an introduction. Springer, Berlin (1977) Cited on pp. 19, 42, 47, and 49.

Haken, H.: Information and Self-Organization: A Macroscopic Approach to Complex Systems. Springer, Berlin (1999) Cited on p. 35.
Hales, T.C.: The honeycomb conjecture. Discrete and Computational Geometry 25(1), 1–22 (2001) Cited on p. 115.
Hamann, H.: Modeling and investigation of robot swarms. Master's thesis, University of Stuttgart, Germany (2006) Cited on pp. 115 and 117.
Hamann, H., Wörn, H.: An analytical and spatial model of foraging in a swarm of robots. In: Şahin, E., Spears, W.M., Winfield, A.F.T. (eds.) SAB 2006 Ws 2007. LNCS, vol. 4433, pp. 43–55. Springer, Heidelberg (2007a) Cited on pp. 67 and 105.
Hamann, H., Wörn, H.: A space- and time-continuous model of self-organizing robot swarms for design support. In: First IEEE International Conference on Self-Adaptive and Self-Organizing Systems (SASO 2007), Boston, USA, July 9-11, pp. 23–31. IEEE Press, Los Alamitos (2007) Cited on pp. 41, 70, and 89.
Hamann, H., Wörn, H.: Embodied computation. Parallel Processing Letters 17(3), 287–298 (2007c) Cited on pp. 24, 68, and 114.
Hamann, H., Wörn, H.: Aggregating robots compute: An adaptive heuristic for the euclidean steiner tree problem. In: Asada, M., Hallam, J.C.T., Meyer, J.-A., Tani, J. (eds.) SAB 2008. LNCS (LNAI), vol. 5040, pp. 447–456. Springer, Heidelberg (2008a) Cited on p. 114.
Hamann, H., Wörn, H.: A framework of space-time continuous models for algorithm design in swarm robotics. Swarm Intelligence, Special Issue: Swarm Robotics 2(2-4), 209–239 (2008b) Cited on pp. 41, 62, 78, 89, 90, 92, 95, 96, 97, 98, 99, 100, 101, 102, 103, and 104.
Hamann, H., Szymanski, M., Wörn, H.: Orientation in a trail network by exploiting its geometry for swarm robotics. In: Shi, Y., Dorigo, M. (eds.) IEEE Swarm Intelligence Symposium, Honolulu, USA, April 1-5, pp. 310–315. IEEE Press, Los Alamitos (2007) Cited on pp. 9 and 15.
Hamann, H., Wörn, H., Crailsheim, K., Schmickl, T.: Spatial macroscopic models of a bio-inspired robotic swarm algorithm. In: IEEE/RSJ 2008 International Conference on Intelligent RObots and Systems (IROS 2008), pp. 1415–1420. IEEE Press, Los Alamitos (2008) Cited on pp. 33, 78, 80, 86, 87, and 88.
Hamann, H., Schmickl, T., Wörn, H., Crailsheim, K.: Analysis of emergent symmetry breaking in collective decision making. Neural Computing & Applications (in press, 2010) Cited on p. 78.
Hänel, D.: Molekulare Gasdynamik – Einführung in die kinetische Theorie der Gase und Lattice-Boltzmann-Methoden. Springer, Berlin (2004) Cited on p. 31.
Hauge, E.H., Martin-Löf, A.: Fluctuating hydrodynamics and brownian motion. J. Stat. Phys. 7(3), 259–281 (1973) Cited on p. 43.
Helbing, D., Schweitzer, F., Keltsch, J., Molnar, P.: Active walker model for the formation of human and animal trail systems. Physical Review E 56(3), 2527–2539 (1997) Cited on p. 30.
Helbing, D., Farkas, I.J., Vicsek, T.: Simulating dynamical features of escape panic. Nature 407, 487–490 (2000) Cited on p. 70.
Higham, N.J.: Accuracy and Stability of Numerical Algorithms. Society for Industrial and Applied Mathematics (2002) Cited on p. 55.
Daniel Hillis, W.: The Pattern On The Stone: The Simple Ideas That Make Computers Work. Basic Books, New York (October 1999) Cited on p. 38.
Hofstadter, D.R.: Gödel, Escher, Bach: An Eternal Golden Braid. Basic Books, New York (1979) Cited on p. 69.

Hogg, T.: Coordinating microscopic robots in viscous fluids. Autonomous Agents and Mutli-Agent Systems 14(3), 271–305 (2006) Cited on p. 30.
Holland, J.H.: Adaptation in Natural and Artificial Systems. MIT Press, Cambridge (1992) Cited on p. 75.
Holland, J.H.: Emergence - From Chaos to Order. Oxford University Press, New York (1998) Cited on p. 8.
H'walisz, L., Jung, P., Hänggi, P., Talkner, P., Schimansky-Geier, L.: Colored noise driven systems with inertia. Zeitschrift für Physik B Condensed Matter 77, 471 (1989) Cited on p. 43.
Hwang, F.K., Richards, D.S., Winter, P.: The Steiner Tree Problem. North-Holland, Amsterdam (1992) Cited on p. 114.
Iida, F.: Cheap design approach to adaptive behavior: Walking and sensing through body dynamics. In: International Symposium on Adaptive Motion of Animals and Machines (2005) Cited on p. 15.
Itô, K.: On stochastic differential equations. Memoirs of the American Mathematical Society 4(1), 200–223 (1951) Cited on p. 43.
Jasmine. Swarm robot – project website (2008), http://www.swarmrobot.org/ Cited on pp. 15, 63, and 78.
Johnson, S.: Emergence: The Connected Lives of Ants, Brains, Cities, and Software. Scribner (2001) Cited on pp. 2 and 20.
Jones, C., Matarić, M.J.: Adaptive division of labor in large-scale minimalist multi-robot systems. In: Proceeding of IEEE/RSJ International Conference on Intelligent Robots and Systems (IROS'03), Las Vegas, NV, USA, October 27-31, vol. 2, pp. 1969–1974. IEEE Press, Los Alamitos (2003) Cited on pp. 14 and 15.
van Kampen, N.G.: Stochastic processes in physics and chemistry. North-Holland, Amsterdam (1981) Cited on pp. 47 and 52.
Kennedy, J., Eberhart, R.C.: Swarm Intelligence. Morgan Kaufmann, San Francisco (2001) Cited on p. 9.
Kennedy, J., Eberhart, R.C.: Particle swarm optimization. In: Proceedings of IEEE International Conference on Neural Networks, p. 4 (1995) Cited on p. 11.
Khatib, O.: Real-time obstacle avoidance for manipulators and mobile robots. International Journal of Robotics Research 5, 90–98 (1986) Cited on p. 105.
Klein, J.: Continuous 3D agent-based simulations in the breve simulation environment. In: Proceedings of NAACSOS Conference (North American Association for Computational, Social, and Organizational Sciences) (2003) Cited on p. 105.
Kolb, M., Herrmann, H.J.: The sol-gel transition modelled by irreversible aggregation of clusters. J. Physics A 18(8), L435–L441 (1985) Cited on p. 114.
Kolmogorov, A.N.: Über die analytischen Methoden in der Wahrscheinlichkeitsrechnung. Mathematische Annalen 104(1), 415–458 (1931) Cited on p. 42.
Kornienko, S., Kornienko, O., Levi, P.: IR-based communication and perception in microrobotic swarms. In: Proceedings of the IEEE/RSJ International conference on intelligent robots and systems (IROS 2005), Edmonton, Canada. IEEE Press, Los Alamitos (2005a) Cited on pp. 15 and 78.
Kornienko, S., Kornienko, O., Levi, P.: Swarm embodiment – a new way for deriving emergent behavior in artificial swarms. In: Levi, P., Schanz, M., Lafrenz, R., Avrutin, V. (eds.) Autonome Mobile Systeme, pp. 25–32 (2005b) Cited on pp. 29 and 38.
Kottege, N., Zimmer, U.R.: Acoustical methods for azimuth, range and heading estimation in underwater swarms. In: Proceedings of Acoustics 2008, Paris, France (June 2008) Cited on p. 15.

Ronald Kube, C., Bonabeau, E.: Cooperative transport by ants and robots. Robotics and Autonomous Systems 30, 85–101 (2000) Cited on p. 14.

Kubík, A.: On emergence in evolutionary multiagent systems. In: Kelemen, J., Sosík, P. (eds.) ECAL 2001. LNCS (LNAI), vol. 2159, pp. 326–337. Springer, Heidelberg (2001) Cited on p. 8.

Kubík, A.: Toward a formalization of emergence. Artificial Life 9, 41–65 (2003) Cited on pp. 8 and 22.

Kuramoto, Y.: Chemical Oscillations, Waves, and Turbulence. Springer, Heidelberg (1984) Cited on p. 31.

Langevin, P.: Sur la théorie du mouvement brownien. Comptes-rendus de l'Académie des Sciences 146, 530–532 (1908) Cited on pp. 42 and 43.

Langton, C.G. (ed.): Proceedings of the Interdisciplinary Workshop on the Synthesis and Simulation of Living Systems (ALIFE 1987), Los Alamos, NM, USA, September 1987. Santa Fe Institute Studies in the Sciences of Complexity, vol. 6. Addison-Wesley, Reading (1987a) Cited on p. 17.

Langton, C.G. (ed.): Artificial Life: Proceedings of an Interdisciplinary Workshop on the Synthesis and Simulation of Living Systems. Addison-Wesley, Reading (1989b) Cited on p. 28.

LaValle, S.M., Kuffner, J.J.: Rapidly-exploring random trees: Progress and prospects. In: Donald, B.R., Lynch, K.M., Rus, D. (eds.) Algorithmic and Computational Robotics: New Directions, Wellesley, MA, USA, pp. 293–308. A. K. Peters (2001) Cited on p. 114.

Leiber, T.: Vom mechanistischen Weltbild zur Selbstorganisation des Lebens, Alber, Freiburg (Breisgau), Munich. Alber-Reihe Thesen, vol. 6 (2000) Cited on p. 32.

Lemons, D.S., Gythiel, A.: Paul Langevin's 1908 paper "On the theory of Brownian motion Sur la théorie du mouvement brownien. Comptes-rendus de l'Académie des Sciences (Paris) 146, 530–533 (1908); American Journal of Physics 65(11), 1079–1081 (Nonember 1997) Cited on p. 42.

Lerman, K.: A model of adaptation in collaborative multi-agent systems. Adaptive Behavior 12(3-4), 187–198 (2004) Cited on p. 28.

Lerman, K., Galstyan, A.: Mathematical model of foraging in a group of robots: Effect of interference. Autonomous Robots 13, 127–141 (2002) Cited on p. 28.

Lerman, K., Martinoli, A., Galstyan, A.: A review of probabilistic macroscopic models for swarm robotic systems. In: Şahin, E., Spears, W.M. (eds.) Swarm Robotics 2004. LNCS, vol. 3342, pp. 143–152. Springer, Heidelberg (2005) Cited on pp. 28 and 67.

Lerman, K., Jones, C., Galstyan, A., Matarić, M.J.: Analysis of dynamic task allocation in multi-robot systems. Int. J. of Robotics Research 25(3), 225–241 (2006) Cited on pp. 15 and 28.

Lewes, G.H.: Problems of Life and Mind (First Series), vol. 2. Trench and Turber & Co., London (1875) Cited on p. 20.

Lighthill, M.J., Whitham, G.B.: On kinematic waves. ii. a theory of traffic flow on long crowded roads. Proceedings of the Royal Society of London A229(1178), 317–345 (1955) Cited on p. 33.

Litus, Y., Zebrowski, P., Vaughan, R.: Energy-efficient multi-robot rendezvous: Parallel solutions by embodied approximation. In: Workshop on Algorithmic equivalencies between biological and robotic swarms, Atlanta, USA (June 2007) Cited on p. 69.

Llinás, R.: I of the Vortex: From Neurons to Self. MIT Press, Cambridge (2002) Cited on p. 24.

Martinoli, A.: Swarm Intelligence in Autonomous Collective Robotics: From Tools to the Analysis and Synthesis of Distributed Control Strategies. PhD thesis, Ecole Polytechnique Fédérale de Lausanne (1999) Cited on pp. 8, 28, and 60.

Martinoli, A., Easton, K., Agassounon, W.: Modeling swarm robotic systems: A case study in collaborative distributed manipulation. Int. Journal of Robotics Research 23(4), 415–436 (2004) Cited on p. 28.

Matarić, M.J.: Reinforcement learning in the multi-robot domain. Autonomous Robots 4(1), 73–83 (1997) Cited on p. 39.

McCanne, S., Floyd, S., Fall, K., Varadhan, K., et al: Network simulator - ns-2 (1997), `http://www-mash.cs.berkeley.edu/ns/` Cited on pp. 16 and 66.

Mill, J.S.: A System of Logic: Ratiocinative and Inductive. John W. Parker and Son, London (1843) Cited on p. 20.

Millonas, M.M.: Swarms, phase transitions, and collective intelligence. In: Langton, C.G. (ed.) Artificial Life III, Addison-Wesley, Reading (1994) Cited on p. 9.

Milutinovic, D., Lima, P.: Cells and Robots: Modeling and Control of Large-Size Agent Populations. Springer, Heidelberg (2007) Cited on p. 29.

Mitchell, M.: Self-awareness and control in decentralized systems. In: Working Papers of the AAAI 2005 Spring Symposium on Metacognition in Computation. AAAI Press, Menlo Park (2005) Cited on p. 18.

Mitchell, M.: Complex systems: Network thinking. Artificial Intelligence 170(18), 1194–1212 (2006) Cited on p. 18.

Mogilner, A., Edelstein-Keshet, L.: A non-local model for a swarm. Journal of Mathematical Biology 38(6), 534–570 (1999) Cited on p. 32.

Mondada, F., Gambardella, L.M., Floreano, D., Nolfi, S., Deneubourg, J.-L., Dorigo, M.: The cooperation of swarm-bots: Physical interactions in collective robotics. IEEE Robotics & Automation Magazine 12(2), 21–28 (2005) Cited on p. 15.

Moran, D.: Introduction to Phenomenology. Routledge, New York (2000) Cited on p. 6.

Murray, J.D.: Mathematical Biology II. Springer, Heidelberg (2003) Cited on p. 32.

Nagel, K., Schreckenberg, M.: A cellular automaton model for freeway traffic. Journal de Physique I 2, 2221–2229 (1992) Cited on p. 33.

De Nardi, R., Holland, O.E.: Ultraswarm: A further step towards a flock of miniature helicopters. In: Şahin, E., Spears, W.M., Winfield, A.F.T. (eds.) SAB 2006 Ws 2007. LNCS, vol. 4433, pp. 116–128. Springer, Heidelberg (2007) Cited on p. 15.

Nembrini, J., Winfield, A.F.T.: Emergent swarm morphology control of wireless networked mobile robots. Journal of Adaptive Behaviour (in press, 2010) Cited on p. 100.

Nembrini, J., Winfield, A.F.T., Melhuish, C.: Minimalist coherent swarming of wireless networked autonomous mobile robots. In: Hallam, B., Floreano, D., Hallam, J., Hayes, G., Meyer, J.-A. (eds.) Proceedings of the seventh international conference on simulation of adaptive behavior on From animals to animats, pp. 373–382. MIT Press, Cambridge (2002) Cited on pp. 15, 29, and 100.

von Neumann, J.: The Theory of Self-Reproducing Automata. In: Burks, A. (ed.), University of Illinois Press, Champaign (1966) Cited on p. 27.

Nicolis, G., Prigogine, I.: Self-organization in nonequilibrium systems. Wiley, New York (1977) Cited on p. 18.

Okubo, A.: Dynamical aspects of animal grouping: Swarms, schools, flocks, and herds. Advances in Biophysics 22, 1–94 (1986) Cited on p. 32.

Okubo, A., Levin, S.A.: Diffusion and Ecological Problems: Modern Perspectives. Springer, Berlin (2001) Cited on p. 32.

Panait, L., Luke, S.: Ant foraging revisited. In: Pollack, J., Bedau, M., Husbands, P., Ikegami, T. (eds.) ALife IX Proceedings, MIT Press, Cambridge (2004) Cited on p. 106.
Panait, L., Luke, S.: Cooperative multi-agent learning: The state of the art. Autonomous Agents and Multi-Agent Systems 11(3), 387–434 (2005) Cited on p. 39.
Van Dyke Parunak, H., Brueckner, S.: Entropy and self-organization in multi-agent systems. In: AGENTS 2001: Proceedings of the fifth international conference on Autonomous agents, pp. 124–130. ACM Press, New York (2001) Cited on p. 19.
Van Dyke Parunak, H., Brueckner, S.A., Sauter, J.A., Matthews, R.: Global convergence of local agent behaviors. In: Proceedings of the Fourth International Joint Conference on Autonomous Agents and Multi-Agent Systems (AAMAS 2005), Utrecht, Netherlands, July 2005, pp. 305–312 (2005) Cited on p. 35.
Payton, D., Daily, M., Estowski, R., Howard, M., Lee, C.: Pheromone robotics. Autonomous Robots 11(3), 319–324 (2001) Cited on pp. 9, 15, 24, 68, 69, and 105.
Peak, D., West, J.D., Messinger, S.M., Mott, K.A.: Evidence for complex, collective dynamics and emergent, distributed computation in plants. Proceedings of the National Academy of Science 101(4), 918–922 (2004) Cited on pp. 70 and 75.
Pfeifer, R., Bongard, J.C.: How the body shapes the way we think – A new view of intelligence. MIT Press, Cambridge (2006) Cited on p. 15.
Planck, M.: Über einen Satz der statistischen Dynamik and seine Erweiterung in der Quantentheorie. Sitzungsberichte der Preußischen Akademie der Wissenschaften 24, 324–341 (1917) Cited on p. 42.
Prigogine, I.: Introduction to thermodynamics of irreversible processes. Wiley, Chichester (1967) Cited on p. 18.
Prigogine, I.: Structure, dissipation and life. Theoretical Physics and Biology, 23–52 (1969) Cited on p. 18.
Prigogine, I.: The End of Certainty: Time, Chaos, and The New Laws of Nature. Free Press, New York (1997) Cited on pp. 22, 24, 30, and 64.
Resnick, M.: Turtles, Termites, and Traffic Jams. MIT Press, Cambridge (1994) Cited on p. 22.
Reynolds, C.W.: Flocks, herds, and schools: A distributed behavioral model. Computer Graphics 21(4), 25–34 (1987) Cited on p. 10.
Risken, H.: The Fokker–Planck Equation. Springer, Berlin (1984) Cited on pp. 42 and 52.
Robins, G., Zelikovsky, A.: Improved Steiner Tree approximation in graphs. In: Proceedings of the 11th ACM-SIAM Symposium on Discrete Algorithms, pp. 770–779 (2000) Cited on p. 114.
Russell, S.J., Norvig, P.: Artificial Intelligence: A Modern Approach. Prentice Hall, Englewood Cliffs (1995) Cited on p. 5.
Scalapino, D.J., Sears, M., Ferrell, R.A.: Statistical mechanics of one-dimensional ginzburg-landau fields. Phys. Rev. B 6(9), 3409–3416 (1972) Cited on p. 19.
Schillo, M., Fischer, K., Klein, C.T.: The micro-macro link in DAI and sociology. In: Moss, S., Davidsson, P. (eds.) MABS 2000. LNCS (LNAI), vol. 1979, pp. 303–317. Springer, Heidelberg (2000) Cited on pp. 9 and 17.
Schmickl, T., Crailsheim, K.: Trophallaxis among swarm-robots: A biologically inspired strategy for swarm robotics. In: Proceedings of the 1st IEEE/RAS-EMBS International Conference on Biomedical Robotics and Biomechanotronics (2006) Cited on pp. 15, 33, and 89.
Schmickl, T., Crailsheim, K.: Trophallaxis within a robotic swarm: bio-inspired communication among robots in a swarm. Autonomous Robots 25(1-2), 171–188 (2008) Cited on p. 89.

Schmickl, T., Möslinger, C., Crailsheim, K.: Collective perception in a robot swarm. In: Şahin, E., Spears, W.M., Winfield, A.F.T. (eds.) SAB 2006 Ws 2007. LNCS, vol. 4433, pp. 144–157. Springer, Heidelberg (2007a) Cited on pp. 33, 70, 89, 90, 96, 97, and 105.
Schmickl, T., Möslinger, C., Thenius, R., Crailsheim, K.: Individual adaptation allows collective path-finding in a robotic swarm. International Journal of Factory Automation, Robotics and Soft Computing, 102–108 (2007b) Cited on p. 89.
Schmickl, T., Möslinger, C., Thenius, R., Crailsheim, K.: Bio-inspired navigation of autonomous robots in heterogenous environments. International Journal of Factory Automation, Robotics and Soft Computing 3, 164–170 (2007c) Cited on pp. 33 and 89.
Schmickl, T., Thenius, R., Möslinger, C., Radspieler, G., Kernbach, S., Crailsheim, K.: Get in touch: Cooperative decision making based on robot-to-robot collisions. Autonomous Agents and Multi-Agent Systems (August 2008) Cited on pp. 77, 79, and 80.
Schmickl, T., Hamann, H., Wörn, H., Crailsheim, K.: Two different approaches to a macroscopic model of a bio-inspired robotic swarm. Robotics and Autonomous Systems 57(9), 913–921 (2009) Cited on pp. 78, 84, and 85.
Schumacher, R.: Book review: Achim Stephan: Emergenz. Von der Unvorhersagbarkeit zur Selbstorganisation. European Journal of Philosophy 10(3), 415–419 (2002) Cited on p. 21.
Schweitzer, F.: Brownian agent models for swarm and chemotactic interaction. In: Polani, D., Kim, J., Martinetz, T. (eds.) Fifth German Workshop on Artificial Life. Abstracting and Synthesizing the Principles of Living Systems, pp. 181–190. Akademische Verlagsgesellschaft Aka (2002) Cited on p. 30.
Schweitzer, F.: Brownian Agents and Active Particles. In: On the Emergence of Complex Behavior in the Natural and Social Sciences. Springer, Berlin (2003) Cited on pp. 30, 42, 59, and 70.
Schweitzer, F., Schimansky-Geier, L.: Clustering of "active" walkers in a two-component system. Physica A 206, 359–379 (1994) Cited on p. 43.
Schweitzer, F., Lao, K., Family, F.: Active random walkers simulate trunk trail formation by ants. BioSystems 41 (1997) Cited on p. 30.
Seyfried, J., Szymanski, M., Bender, N., Estaña, R., Thiel, M., Wörn, H.: The I-SWARM project: Intelligent small world autonomous robots for micro-manipulation. In: Şahin, E., Spears, W.M. (eds.) Swarm Robotics 2004. LNCS, vol. 3342, pp. 70–83. Springer, Heidelberg (2005) Cited on p. 12.
Sharkey, A.J.C.: Swarm robotics and minimalism. Connection Science 19(3), 245–260 (2007) Cited on p. 14.
von Smoluchowski, M.: Zur kinetischen Theorie der Brownschen Molekularbewegung und der Suspensionen. Annalen der Physik 21, 756–780 (1906) Cited on p. 42.
Soysal, O., Şahin, E.: A macroscopic model for self-organized aggregation in swarm robotic systems. In: Şahin, E., Spears, W.M., Winfield, A.F.T. (eds.) SAB 2006 Ws 2007. LNCS, vol. 4433, pp. 27–42. Springer, Heidelberg (2007) Cited on p. 29.
Spears, W.M., Gordon, D.F.: Using artificial physics to control agents. In: IEEE International Conference on Information, Intelligence, and Systems. IEEE Press, Los Alamitos (1999) Cited on pp. 15 and 105.
Stephan, A.: Emergenz: Von der Unvorhersagbarkeit zur Selbstorganisation. Dresden University Press, Dresden (1999) Cited on p. 20.
Stepney, S.: Embodiment. In: Flower, D., Timmis, J. (eds.) Silico Immunology, ch. 12, pp. 265–288. Springer, Berlin (2007) Cited on p. 24.
Stepney, S., Polack, F., Turner, H.: Engineering emergence. In: CEC 2006: 11th IEEE International Conference on Engineering of Complex Computer Systems, Stanford, CA, USA, August 2006. IEEE Press, Los Alamitos (2006) Cited on pp. 24 and 35.

Sugawara, K., Kazama, T., Watanabe, T.: Foraging behavior of interacting robots with virtual pheromone. In: Proceedings of 2004 IEEE/RSJ International Conference on Intelligent Robots and Systems. IEEE Press, Los Alamitos (2004) Cited on pp. 9 and 105.

Szymanski, M., Wörn, H.: JaMOS – A MDL2e based operating system for swarm micro robotics. In: IEEE Swarm Intelligence Symposium, Honolulu, USA, April 1-5. IEEE Press, Los Alamitos (2007) Cited on pp. 15 and 78.

Theraulaz, G., Bonabeau, E., Nicolis, S.C., Solé, R.V., Fourcassié, V., Blanco, S., Fournier, R., Joly, J.-L., Fernández, P., Grimal, A., Dalle, P., Deneubourg, J.-L.: Spatial patterns in ant colonies. Proc. Natl. Acad. Sci. USA 99(15), 9645–9649 (2002) Cited on p. 22.

Theraulaz, G., Gautrais, J., Camazine, S., Deneubourg, J.-L.: The formation of spatial patterns in social insects: from simple behaviours to complex structures. Philosophical Transactions of the Royal Society of London. Series A 361, 1263–1282 (2003) Cited on p. 32.

Witten Jr., T.A., Sander, L.M.: Diffusion-limited aggregation, a kinetic critical phenomenon. Phys. Rev. Lett. 19, 1400–1403 (1981) Cited on pp. 7, 114, and 115.

Trautz, M.: Das Gesetz der Reaktionsgeschwindigkeit und der Gleichgewichte in Gasen. Bestätigung der Additivität von Cv-3/2R. Neue Bestimmung der Integrationskonstanten und der Moleküldurchmesser. Zeitschrift für anorganische und allgemeine Chemie 96(1), 1–28 (1916) Cited on pp. 31 and 82.

Trianni, V.: Evolutionary Swarm Robotics – Evolving Self-Organising Behaviours in Groups of Autonomous Robots. Studies in Computational Intelligence, vol. 108. Springer, Berlin (2008) Cited on p. 39.

Turing, A.M.: The chemical basis of morphogenesis. Philosophical Transactions of the Royal Society of London. Series B, Biological Sciences B237(641), 37–72 (1952) Cited on p. 31.

Vaughan, R.T., Stoy, K., Sukhatme, G.S., Matarić, M.J.: Whistling in the dark: Cooperative trail following in uncertain localization space. In: Autonomous Agents and Multi-Agent Systems, Barcelona, Spain (2000) Cited on p. 15.

Vetulani, Z.: Decision making for a robocup multi-agent system. In: Proceedings of the Third International Workshop on Robot Motion and Control (RoMoCo 2002), November 2002, pp. 193–200 (2002) Cited on p. 17.

Vicsek, T., Czirok, A., Ben-Jacob, E., Cohen, I., Shochet, O.: Novel type of phase transition in a system of self-driven particles. Physical Review Letters 6(75), 1226–1229 (1995) Cited on p. 30.

Warme, D., Winter, P., Zachariasen, M.: Geosteiner homepage (2008), `http://www.diku.dk/geosteiner/` Cited on p. 114.

Weinberg, S.: Reductionism redux. The New York Review of Books (October 5, 1995) Cited on pp. 8 and 23.

Weiss, G.: Multiagent Systems: A Modern Approach to Distributed Artificial Intelligence. MIT Press, Cambridge (1999) Cited on p. 16.

Wilson, M., Melhuish, C., Sendova-Franks, A.B., Scholes, S.: Algorithms for building annular structures with minimalist robots inspired by brood sorting in ant colonies. Autonomous Robots 17, 115–136 (2004) Cited on p. 14.

Winfield, A.F.T., Sav, J., Fernández-Gago, M.-C., Dixon, C., Fisher, M.: On formal specification of emergent behaviours in swarm robotic systems. International Journal of Advanced Robotic Systems 2(4), 363–370 (2005) Cited on p. 29.

De Wolf, T., Holvoet, T.: Emergence versus self-organisation: Different concepts but promising when combined. In: Brueckner, S.A., Di Marzo Serugendo, G., Karageorgos, A., Nagpal, R. (eds.) ESOA 2005. LNCS (LNAI), vol. 3464, pp. 1–15. Springer, Heidelberg (2005) Cited on pp. 7 and 8.

Woolridge, M.: Introduction to Multiagent Systems. John Wiley & Sons, Inc., New York (2001) Cited on p. 16.

WordNet 3.0, Princeton University (2007), `http://wordnet.princeton.edu/perl/webwn?s=computation` Cited on p. 69.

Wörn, H., Seyfried, J., Fahlbusch, S., Bürkle, A., Schmöckel, F.: Flexible microrobots for micro assembly tasks. In: International Symposium on Micromechatronics and Human Science (HMS 2000), Nagoya, pp. 22–25 (2000) Cited on p. 11.

Yamins, D.: Towards a theory of "local to global" in distributed multi-agent systems. In: Proceedings of the fourth international joint conference on Autonomous agents and multiagent systems (AAMS 2005), Utrecht, Netherlands, pp. 183–190 (2005) Cited on pp. 18 and 35.

Yamins, D.: A Theory of Local-to-Global Algorithms for One-Dimensional Spatial Multi-Agent Systems. PhD thesis, Harvard University, Cambridge (November 2007) Cited on pp. 18, 35, and 57.

Yamins, D., Nagpal, R.: Automated global-to-local programming in 1-d spatial multi-agent systems. In: Padgham, L., Parkes, D.C., Müller, J.P., Parsons, S. (eds.) Proc. of 7th Int. Conf. on Autonomous Agents and Multiagent Systems (AAMAS 2008), Estoril, Portugal (May 2008) Cited on pp. 18 and 35.

Zachariasen, M., Winter, P.: Concatenation-based greedy heuristics for the Steiner tree problem in the Euclidean plane. Algorithmica 25, 418–437 (1999) Cited on pp. 114 and 121.

Index

Cognitive Systems Monographs

Edited by R. Dillmann, Y. Nakamura, S. Schaal and D. Vernon

Vol. 1: Arena, P.; Patanè, L. (Eds.):
Spatial Temporal Patterns for
Action-Oriented Perception
in Roving Robots
425 p. 2009 [978-3-540-88463-7]

Vol. 2: Ivancevic, T.T.; Jovanovic, B.;
Djukic, S.; Djukic, M.; Markovic, S.:
Complex Sports Biodynamics
326 p. 2009 [978-3-540-89970-9]

Vol. 3: Magnani, L.:
Abductive Cognition
534 p. 2009 [978-3-642-03630-9]

Vol. 4: Azad, P.:
Visual Perception for Manipulation
and Imitation in Humanoid Robots
270 p. 2009 [978-3-642-04228-7]

Vol. 5: de de Aguiar, E.:
Animation and Performance Capture
Using Digitized Models
168 p. 2009 [978-3-642-10315-5]

Vol. 6: Ritter, H.; Sagerer, G.;
Dillmann, R.; Buss, M.:
Human Centered Robot Systems
216 p. 2009 [978-3-642-10402-2]

Vol. 7: Levi, P.; Kernbach, S. (Eds.):
Symbiotic Multi-Robot Organisms
467 p. 2010 [978-3-642-11691-9]

Vol. 8: Christensen, H.I.;
Kruijff, G.-J.M.; Wyatt, J.L. (Eds.):
Cognitive Systems
491 p. 2010 [978-3-642-11693-3]

Vol. 9: Hamann, H.:
Space-Time Continuous Models of Swarm
Robotic Systems
147 p. 2010 [978-3-642-13376-3]

GPSR Compliance

The European Union's (EU) General Product Safety Regulation (GPSR) is a set of rules that requires consumer products to be safe and our obligations to ensure this.

If you have any concerns about our products, you can contact us on ProductSafety@springernature.com

In case Publisher is established outside the EU, the EU authorized representative is:

Springer Nature Customer Service Center GmbH
Europaplatz 3
69115 Heidelberg, Germany

Batch number: 10614853

Printed by Printforce, the Netherlands